PROTEIN BIOSYNTHESIS IN NONBACTERIAL SYSTEMS

METHODS IN MOLECULAR BIOLOGY

Edited by

ALLEN I. LASKIN
ESSO Research and Engineering Company
Linden, New Jersey

JEROLD A. LAST
National Academy of Sciences
Washington, D.C.

VOLUME 1: Protein Biosynthesis in Bacterial Systems, *edited by Jerold A. Last and Allen I. Laskin*

VOLUME 2: Protein Biosynthesis in Nonbacterial Systems, *edited by Jerold A. Last and Allen I. Laskin*

VOLUMES IN PREPARATION

VOLUME 3: Nucleic Acids, *edited by Allen I. Laskin and Jerold A. Last*

VOLUME 4: Subcellular Particles, Structures, and Organelles, *edited by Allen I. Laskin and Jerold A. Last*

PROTEIN BIOSYNTHESIS IN NONBACTERIAL SYSTEMS

EDITED BY

Jerold A. Last and Allen I. Laskin

National Academy of Sciences
Washington, D.C.

ESSO Research and Engineering Company
Linden, New Jersey

MARCEL DEKKER, INC. New York 1972

MARCEL DEKKER, INC.
95 Madison Avenue, New York, New York 10016

LIBRARY OF CONGRESS CATALOG CARD NUMBER: 78-189798
ISBN: 0-8247-1397-4

PRINTED IN THE UNITED STATES OF AMERICA

PREFACE

This book, Volume 2 in the series entitled *Methods in Molecular Biology*, continues the format begun in Volume 1 of providing small, relatively inexpensive, topically organized volumes, presenting critically written descriptions of methods used in a particular area of the field. It is expected that these books will be particularly useful to new workers entering the field, to graduate students beginning a problem, to new technicians, etc.

The authors were asked to write descriptions of the methods used in a particular area, and whenever appropriate, to discuss such things as: *why* a particular approach was taken, *why* a particular reagent was used, what alternatives are feasible and acceptable, what to do "if things go wrong," etc.

This volume describes, from a practical viewpoint, methods used for research in the area of *in vitro* protein biosynthesis in nonbacterial systems, and is intended to continue the type of presentation in Volume 1, which dealt with *in vitro* protein biosynthesis in bacterial systems. The methods are critically explored in terms of the reasons for their selection, and possible alternative techniques are considered in the discussion.

In treating the vast range of "nonbacterial systems," the book describes the preparation of cell-free systems from mammalian tissues such as muscle, reticulocytes, spleen, and brain. There is a chapter on preparation of active extracts from various tissue culture cells, and another on extracts prepared from tissue culture cells in which mammalian viruses have been grown. The preparation of a specific inhibitor of viral protein synthesis, interferon, is detailed. Other sources described for cell-free systems active in protein synthesis include plants (wheat germ) and unicellular animals (paramecia). The chapter on paramecia also introduces some immunological techniques useful for the study of protein biosynthesis. The remainder of the volume concerns purified components of mammalian protein-synthesizing machinery: initiation factors from reticulocytes and messenger RNA from reticulocytes.

Washington, D. C. Jerold A. Last
Linden, New Jersey Allen I. Laskin

CONTRIBUTORS TO THIS VOLUME

W. FRENCH ANDERSON, Section on Human Biochemistry, National Heart and Lung Institute, National Institutes of Health, Bethesda, Maryland

RALPH B. ARLINGHAUS, Section on Environmental Biology, M. D. Anderson Hospital, University of Texas Medical Center, Houston, Texas

RICHARD ASCIONE, Animal Disease and Parasite Research Division, Plum Island Animal Disease Laboratory, Greenport, New York.

JAMES J. CASTLES, Department of Medicine, University of Chicago, Chicago, Illinois

HILTON B. LEVY, Laboratory of Viral Diseases, National Institute of Allergy and Infectious Diseases, Bethesda, Maryland

JERRY B. LINGREL, Department of Biological Chemistry, College of Medicine, University of Cincinnati, Cincinnati, Ohio

ABRAHAM MARCUS, The Institute for Cancer Research, Fox Chase, Philadelphia, Pennsylvania

THOMAS C. MERIGAN, Division of Infectious Diseases, Stanford University School of Medicine, Stanford, California

PHILIP M. PRICHARD, Section on Human Biochemistry, National Heart and Lung Institute, National Institutes of Health, Bethesda, Maryland

DAVID A. SHAFRITZ, Section on Human Biochemistry, National Heart and Lung Institute, National Institutes of Health, Bethesda, Maryland

JOHN SOMMERVILLE, Department of Zoology, University of St. Andrews, St. Andrews, Fife, Scotland

NORMAN TALAL, Department of Health, Education, and Welfare, Public Health Service, National Institutes of Health, Bethesda, Maryland

GEORGE F. VANDE WOUDE, Plum Island Animal Disease Laboratory, Veterinary Sciences Research Division, Agricultural Research Service, U.S. Department of Agriculture, Greenport, New York

IRA G. WOOL, Department of Physiology, University of Chicago, Chicago, Illinois

CLAIRE E. ZOMZELY-NEURATH, Department of Biochemistry, Roche Institute of Molecular Biology, Nutley, New Jersey

CONTENTS

Preface iii

Contributors to This Volume v

Chapter 1. PREPARATION OF SKELETAL MUSCLE RIBOSOMES AND ASSAY OF PROTEIN SYNTHESIS

James J. Castles and Ira G. Wool

I. Introduction 1

II. Animals. 2

III. Preparation of Skeletal Muscle Ribosomes 2

IV. Preparation of Ribosomal Subunits. 7

V. Assay of Protein Synthesis by Muscle Ribosomes 16

References 26

Chapter 2. ISOLATION OF MAMMALIAN CELL POLYRIBOSOMES

Ralph B. Arlinghaus and Richard Ascione

I. Introduction 30

II. Isolation of Rabbit Reticulocyte Polyribosomes 35

III. Preparation of Ribosomes from Reticulocytes Free of Transfer Factors .37

IV. Preparation of Polyribosomes from a Cultured Mammalian Cell 40

V. Preparation of Hepatic Polyribosomes 44

VI. Amino Acid Incorporation System. 51

VII. Preparation of Enzymes for Protein Synthesis. 52

VIII. Sucrose Density Gradient Centrifugation 53

Abbreviations 55

References. 55

Chapter 3. TISSUE CULTURE POLYRIBOSOMAL SYSTEMS

Richard Ascione, Ralph B. Arlinghaus, and George F. Vande Woude

I. Introduction. 60

II. Effect of RNase Inhibitors on Polyribosome Isolation 66

III. Effect of Increasing Concentrations of Dextran Sulfate on Isolation of Polyribosome Preparations . .67

IV. Effect of Cations and Detergents on the Yields of Polyribosomes 70

V. Incubation Conditions of Tissue Culture Cells and Polyribosomes 74

VI. Isolation of Polyribosomes from Mammalian Cells . . .78

VII. Virus-Infected Tissue Culture Cells 83

VIII. Preparation of Membrane-Bound and Free Ribosomes. . .92

IX. Cell-Free Studies with Tissue Culture Polyribosomes 96

X. Requirements for Cell-Free Protein Synthesis. . . . 100

XI. Preparation of Tissue Culture Ribosomal Subunits. . 105

XII. Preparation of Ribosome-Associated I Factors from BHK Polyribosomes 106

XIII. Incorporation Assay with Subunits and I Factor. . . 108

XIV. I-Factor-Dependent Binding of N-Acetyl-L-Phe-tRNA . 109

Abbreviations 111

References. 112

Chapter 4. POLYRIBOSOMES AND CELL-FREE PROTEIN SYNTHESIS IN THE SPLEEN

Norman Talal

I. Introduction 117

II. Preparation of Spleen Polyribosomes. 119

III. Preparation of Membrane-Bound and Free Ribosomes . . 120

IV. Cell-Free Protein Synthesis. 122

V. Binding of Radioactive Poly U to Ribosomes 125

References 126

Chapter 5. PROTEIN SYNTHESIS IN EXTRACTS OF WHEAT EMBRYO

Abraham Marcus

I. Poly U-Dependent Incorporation of Phenylalanine. . . 128

II. Tobacco Mosaic Virus RNA-Dependent Incorporation of Aminoacyl-tRNA into Polypeptides. 135

References 144

Chapter 6. CEREBRAL PROTEIN-SYNTHESIZING SYSTEMS

Claire E. Zomzely-Neurath

I. Introduction 147

II. Preparation of Cerebral Microsomes, Ribosomes, and pH 5 Enzymes 149

III. Preparation of Total and Free Polyribosomes. 159

IV. Cell-Free Amino Acid-Incorporating System. 168

V. Preparation of RNA with Properties of mRNA 176

VI. Preparation of "Stripped" Ribosomes. 183

References 184

Chapter 7. PROTEIN BIOSYNTHESIS IN *Paramecium* WITH SPECIAL REFERENCE TO THE *IN VITRO* SYNTHESIS OF THE CELL SURFACE ANTIGENS

John Sommerville

I. Introduction 190

II. General Procedures in the Preparation of *Paramecium* Cell Fractions. 193

III. Amino Acid Incorporation by *Paramecium* Cell-Free Systems. 205

IV. Detection and Assay of Labeled Surface Antigen Proteins 213

V. Methods of Tracing the Biosynthetic Pathway of Surface Antigen Proteins 221

VI. Other Protozoan Systems. 226

References 227

Chapter 8. PREPARATION AND ASSAY OF HEMOGLOBIN mRNA

Jerry Lingrel

I. Introduction 231

II. Isolation of Hemoglobin mRNA 233

III. Assay of Hemoglobin mRNA 247

References 261

Chapter 9. PREPARATION AND ASSAY OF RETICULOCYTE INITIATION FACTORS

I. Introduction 266

II. Ribosomes and Crude Ribosomal High-Salt Wash Fraction 267

III. Factor M_1. 276

IV. Factor M_2. 282

V. Factor M_3. 285

VI. Selection of Assay Procedure 289

VII. Summary. 290

References 291

Chapter 10. PREPARATION AND MODE OF ACTION OF INTERFERON

Hilton B. Levy and Thomas C. Merigan

I. Introduction 295

II. Interferon Preparation 297

III. Properties of Interferon 300

IV. Purification of Interferons. 303

V. Mode of Action of Interferon 306

Notes. 315

References 315

AUTHOR INDEX. 323

SUBJECT INDEX 333

PROTEIN BIOSYNTHESIS IN NONBACTERIAL SYSTEMS

Chapter 1

PREPARATION OF SKELETAL MUSCLE RIBOSOMES AND ASSAY OF PROTEIN SYNTHESIS

James J. Castles and Ira G. Wool

Departments of Medicine, Biochemistry, and Physiology
University of Chicago
Chicago, Illinois

I. INTRODUCTION 1

II. ANIMALS 2

III. PREPARATION OF SKELETAL MUSCLE RIBOSOMES 2

IV. PREPARATION OF RIBOSOMAL SUBUNITS 7

V. ASSAY OF PROTEIN SYNTHESIS BY MUSCLE RIBOSOMES . . . 16

REFERENCES 26

I. INTRODUCTION

Determination of biochemical reaction mechanisms and analysis of the regulation of protein synthesis are greatly aided by the use of acellular preparations. The number of variables can be reduced and the parameters more easily manipulated. A subcellular preparation capable of catalyzing the synthesis of protein is a necessary antecedent to the purification of the individual components—ribosomes; initiation, translation, and termination factors; and so on—and determination of their mode of function. Korner [1] pioneered the use of isolated ribosomes to analyze the regulation of protein synthesis in animal cells, and the power of the approach has been demonstrated over and over again. We describe the procedures we use to prepare ribosomes and ribosomal subunits from muscle and the methods we use for assaying the synthesis of protein.

II. ANIMALS

Male albino rats (Sprague-Dawley) weighing 150-180 g are used. The use of male rats avoids variations in protein synthesis that may result from cyclic changes in the secretion of ovarian hormones. The animals are kept in the laboratory for at least 2 days and are allowed free access to food and water. It is often recommended that they be starved in order to deplete the tissues of glycogen which can contaminate ribosome preparations. We do not starve the animals since starvation changes muscle ribosome function in a manner similar to diabetes [2] and because muscle ribosomes, unlike liver ribosomes, are not contaminated with glycogen.

III. PREPARATION OF SKELETAL MUSCLE RIBOSOMES

Ribosomes are prepared by a modification [3] of the method described by Florini and Breuer [4]. Rats are killed by decapitation; it is important that the animals be killed quickly, and separation of the carcass from the head (pituitary) is desirable to prevent flooding of the tissues with hormones. The abdominal wall, thigh, and gastrocnemius muscles are removed. These muscles are easily dissected free of adipose tissue and do not have a large content of connective tissue; the latter tissue makes homogenization difficult. The purpose is to obtain as much usable muscle from a single animal as possible so as to maximize the yield and minimize the cost. The physical and functional characteristics of ribosomes from these different muscles are identical [2]. The muscles are pooled and homogenized in 2 vol of buffer containing 50 mM tris-HCl (pH 7.6), 12.5 mM $MgCl_2$, 250 mM KCl, and 250 mM sucrose for 30 sec at a setting of 80 in a VirTis 45 homogenizer.

To avoid degradation of rRNA, the sucrose used in the preparation of ribosomes and subunits should be either commercial, RNase-free grade, or standard reagent grade that has been treated by heating a 60% solution (in water) with 7.2% Norit A to 80°C for 1 hr and passing the mixture through Whatman no. 3 filter paper [5]. The treated sucrose has the additional advantage of having an absorbancy at 260 nm of less than 0.040.

The concentration of KCl (250 mM) is sufficient to solubilize myosin—an ionic strength of 0.3 or greater is required [6]. Because of the insolubility of myosin at lower ionic strengths, ribosomes may be trapped when muscle protein precipitates and the yield of the particles reduced [7]. The molar ratio of potassium to magnesium ions is also important; when it exceeds 35 to 40, ribosomes dissociate into subunits (see below) which may be lost because they do not sediment under the conditions of centrifugation we ordinarily use. There are other ways of fracturing muscle cells—nitrogen cavitation [8] and an Ultra-Turrax homogenizer [9] are examples. Homogenization of muscle and subsequent steps in the preparation of ribosomes are performed at 5°C in order to minimize nuclease and protease activities and to prevent inactivation of heat-labile subcellular components.

The homogenate is centrifuged at low speed (9,000 rpm-13,000 × g for 15 min in a Servall GSA rotor) to remove unfractured cells, nuclei, mitochondria, myofibrils, and membranes. The supernatant is put aside, and the pellet is resuspended in 1 vol of the homogenization buffer and centrifuged to recover ribosomal particles that may have been trapped in the cell debris. The combined supernatants are filtered through glass wool to remove aggregates of fat and glycogen which, because of their low density, do not sediment during the preliminary centrifugation.

Ribosomal particles are collected from the supernatant by high-speed centrifugation (2 hr at 78,000 × g in a Spinco 30 rotor at 30,000 rpm). The ribosome preparation, which is contaminated with cell membrane fragments, sarcoplasmic reticulum, and myosin, must be resolved. To do this, the

78,000 × g pellets obtained from 100 ml of the 13,000 × g supernatant are homogenized (in a size C glass tissue grinder with a motor-driven Teflon pestle, Arthur Thomas Company, no. 4288-B) in 25.5 ml of medium A [50 mM tris-HCl (pH 7.6), 12.5 mM $MgCl_2$, 80 mM KCl] containing 250 mM sucrose, and treated with 1.5 ml of Lubrol WX (10% solution in 10 mM $MgCl_2$) and 3 ml of sodium deoxycholate (DOC) (10% in water). The low ionic strength of medium A causes precipitation of myosin, and the detergents solubilize the membrane fragments and sarcoplasmic reticulum; the purity of the ribosomes is greater when the two detergents are used. After the heavy white precipitate of myosin is removed by centrifugation at 13,000 × g for 15 min (in a Servall GSA rotor at 9000 rpm), the supernatant is diluted with 0.15 vol of medium A; the procedure has been found to reduce the contamination of the final ribosomal pellet with protein—we do not know why. Portions (7 ml) of the solution are layered over 5 ml of medium A containing 500 mM sucrose, and the ribosomes are sedimented by centrifugation for 2.5 hr at 105,000 × g in a Spinco 40 rotor at 40,000 rpm. The sedimentation of ribosomes through the cushion of sucrose removes cytoplasmic proteins adsorbed to the particles. After centrifugation the tubes are drained and wiped dry; the ribosome pellets are stored at -20°C. They can be kept for at least 3 months (and probably considerably longer) with no loss of activity. However, care should be taken to prevent their dehydration. Just prior to experiments the ribosomes are suspended in medium A by gentle manual homogenization (in a size AA glass tissue grinder with a Teflon pestle, Arthur Thomas Company, no. 4288-B); they are incubated for 5 min at 37°C, and aggregates are removed by centrifugation at 3000 × g for 10 min. Concentrated suspensions of ribosomes (greater than 100 A_{260} units per milliliter) can be stored in medium A in liquid nitrogen without loss of activity.

We obtain about 1.8-2.2 mg of ribosomes from the muscles of a 150 to 180 g rat. The particles are 51% RNA and 49% protein. A_{260}/A_{235} and A_{260}/A_{280} are 1.50-1.55 and 1.85-1.95, respectively. Lower ratios indicate contamination with nonribosomal protein; absorption ratios provide a simple and convenient means of judging the relative purity of the particles. The concentration of ribosomes in a suspension can be estimated from the absorption at 260 nm; 1 A_{260} unit is the equivalent of 90 μg of ribosomes. For a more precise determination of the concentrations of ribosomes, a modification [10] of the method of Fleck and Munro [11] can be employed. To 0.1 ml of ribosomes suspended in medium A is added 0.9 ml of 0.3 N KOH (to hydrolyze rRNA); the blank contains 0.1 ml of medium A. After incubation of the sample for 1 hr at 37°C, 2 ml of ice-cold 1 N perchloric acid are added to precipitate ribosomal protein and the mixture is kept at 0°C for 10 min. The solution is then centrifuged at 1600 × g for 5 min at 4°C to remove the precipitated proteins and $KClO_4$; 2 ml of the supernatant are diluted with 2 ml of water for determination of the absorption at 260 nm. The concentration of rRNA is calculated from the following formula: A_{260} units × 218 = micrograms of rRNA per 0.1 ml of original suspension. This procedure eliminates absorbance at 260 nm that may be attributable to proteins adsorbed to ribosomal particles.

Twenty-five is a convenient number of rats for a preparation, and one person can perform the complete operation in 8-10 hr. Modification of the buffers and conditions of the procedure are surely possible; however, we have not studied alternatives extensively. A modification that we use to prepare supernatant factors for assay of protein synthesis is to homogenize the muscle in medium A containing 250 mM sucrose. Because of the low ionic strength, myosin and ribosomes coprecipitate, thus simplifying the preparation of supernatant factors; however, 50% of the ribosomes are lost.

IV. PREPARATION OF RIBOSOMAL SUBUNITS

Muscle ribosomes are dissociated into subunits by high concentrations of potassium; the subunits are capable of recombining to form monomers that synthesize protein in the presence of template RNA [12]. Ribosomes are also dissociated when low concentrations of magnesium are combined with a chelating agent; however, the subunits do not reassociate into 80 S monomers, nor are they active in protein synthesis [3, 13, 14].

Muscle ribosomes are suspended in medium B [50 mM tris-HCl (pH 7.6), 12.5 mM $MgCl_2$, 880 mM kCl, 20 mM β-mercaptoethanol] at a concentration of 50-100 A_{260} units per milliliter. The high potassium concentration in medium B causes dissociation of ribosomes into their component subunits, probably by displacing magnesium which stabilizes the monosome. It is most likely the molar ratio of potassium to magnesium ions that is critical for dissociating eukaryotic ribosomes; Falvey and Staehlin [15] found that mouse liver ribosomes dissociate when the potassium/magnesium ratio exceeds 35 to 40. This is probably also the case for muscle ribosomes (R. Wettenhall, unpublished experiments). The preparation of active subunits requires the presence of a reagent capable of keeping sulfhydryl (SH) groups reduced during dissociation, isolation, and any subsequent steps. We routinely use 20 mM β-mercaptoethanol, but 5 mM dithiothreitol (Cleland's reagent) is equally satisfactory. The suspension is incubated at 37°C for 5 min to decrease formation of subunit aggregates and then clarified by centrifugation for 10 min at 3000 × g. The suspension (1-2 ml, 50-100 A_{260} units) is layered onto a 35-ml (10-30%) linear sucrose gradient containing medium B, and the ribosomal subunits are separated by centrifugation at 95,000 × g (in a Spinco SW-27 rotor at 27,000 rpm) for 3.5-4 hr. The gradients are prepared at room temperature (23°C), and the temperature of centrifugation is 28°C; the temperature is critical for separation of the ribosomal subunits (see below). (See Vol. 1, Chapters 1 and 5, for a detailed discussion of density gradients). After centrifugation the gradients are pumped through an Instruments Specialties Company, Inc. (ISCO) Model D density gradient fractionator equipped with a Model UA-2 analyzer, and fractions corresponding to the 40 and 60 S subunits are collected (Fig. 1A). (The ISCO system is very valuable. It is simple and reliable, and happily eliminates the tedium of manual gradient fractionation; furthermore, the flow cell for the ultraviolet analyzer rarely collects troublesome bubbles.)

Not all the ribosomes dissociate into subunits in 880 mM KCl; particles that were engaged in synthesizing peptides in the cell are relatively resistant to dissociation *in vitro* [16]. The undissociated ribosome monomers have a sedimentation coefficient of 75 (Fig. 1A). Removal of nascent peptides from the resistant monomers (75 S particles) causes them to dissociate in high concentrations of potassium. Nascent peptide is most easily discharged by adding puromycin (to a final concentration of 0.1 mM) to a solution of ribosomes (75-150 A_{260} units per milliliter) in medium B and incubating for 15 min at 37°C. Puromycin reacts with incomplete peptide chains (actually, peptidyl-tRNA) that are in the donor (peptidyl) site on the ribosome; the ribosomal enzyme peptidyl transferase catalyzes the formation of peptidyl-puromycin which is released from the

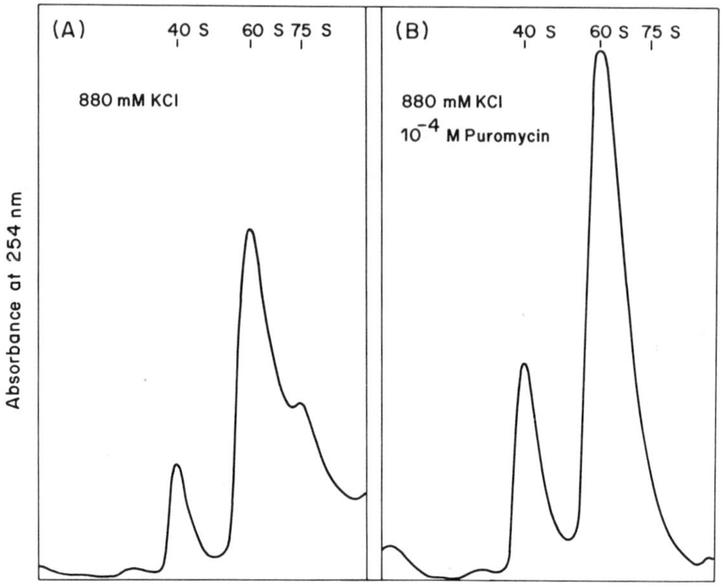

Fig. 1. Dissociation of skeletal muscle ribosomes. (A) Muscle ribosomes in medium B were centrifuged through a sucrose gradient in medium B. (B) Muscle ribosomes were incubated with puromycin in medium B, and then sedimented through a sucrose gradient in medium B.

ribosome. We are not certain whether or not isolated muscle ribosomes also contain peptidyl-tRNA in the acceptor (aminoacyl) site; peptidyl-tRNA in the acceptor site cannot react with puromycin. If there is peptidyl-tRNA in the acceptor site, it is released by the high concentrations of potassium, for none remains on the ribosomes after treatment with puromycin in medium B. The puromycin-treated ribosomes are then layered onto sucrose gradients and centrifuged as described above. Only 40 and 60 S subunits appear in the gradients (Fig. 1B). This is the most efficient and expeditious method of preparing subunits. The procedure has the advantages of increased yield and greater purity; especially, there is less contamination of the 60 S fraction with 40 S subunits. The only reservation is that the antibiotic may modify the ribosomes. Puromycin-stripped ribosomes have a greater tendency to dimerize [15, 17].

The nature and purity of the subunit fractions can be determined from analysis of their RNA. The fractions from preparative gradients are dialyzed overnight at 4°C against 50 mM tris-HCl (pH 7.8) to remove magnesium and potassium which can precipitate sodium dodecyl sulfate (SDS); SDS is then added (to a final concentration of 0.1%) to dissociate protein from rRNA [18]. The sample is incubated at 37°C for 3 min and layered onto a 15-30% linear sucrose gradient containing 50 mM tris-HCl (pH 7.8); centrifugation is at 257,000 × g (in a Spinco SW-65 at 60,000 rpm) for 3 hr at 4°C. The gradient is analyzed with an ISCO gradient fractionator and ultraviolet analyzer. The 40 S fraction (Fig. 1A) contains only 18 S RNA and thus has only 40 S subunits (Fig. 2A); the 60 S fraction (Fig. 1A) contains predominantly 28 S RNA but also a small amount of 18 S RNA (Fig. 2B); this fraction thus consists mainly of 60 S subunits but with some contamination by 40 S particles. (The RNA having sedimentation coefficients between 18 and 28

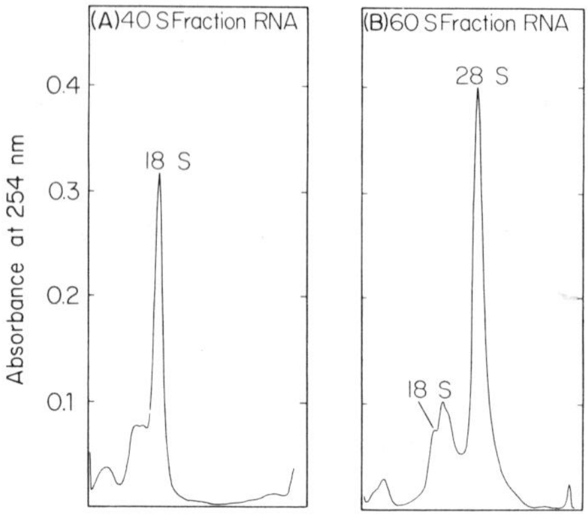

Fig. 2. Analysis of the RNA of the 40 and 60 S fractions from dissociated muscle ribosomes. The RNA present in the 40 S (A) and 60 S (B) fractions isolated from a sucrose gradient similar to that in Figure 1A was determined by centrifugation on SDS gradients.

(Fig. 2B) is thought to arise from the breakdown of 28 S RNA.) The 60 S fraction also contains 5 S RNA which can be detected by electrophoresis on polyacrylamide gels (E. Harris, unpublished data).

Another way of estimating the purity of subunit fractions is centrifugation on analytical sucrose gradients containing medium A. In medium A, subunits reassociate into monomers; consequently, contamination of either subunit fraction is recognized by the presence of 80S ribosomes. The fractions to be analyzed are dialyzed overnight at 4°C against medium A containing 20 mM β-mercaptoethanol and are then layered onto

15-30% linear sucrose gradients in medium A and centrifuged at 257,000 × g (in a Spinco SW-65 at 60,000 rpm) for 40 min at 28°C. The 40 S fraction contains only 40 S subunits (Fig. 3A); the presence of 80 S ribosomes in the 60S fraction (Fig. 3B) indicates that it is contaminated with 40 S subunits (90 S particles are dimers of the 60 S subunit). Generally, the 40 S fraction is less than 5% contaminated with 60 S particles; the 60 S fraction contains 10-20% 40 S particles. When

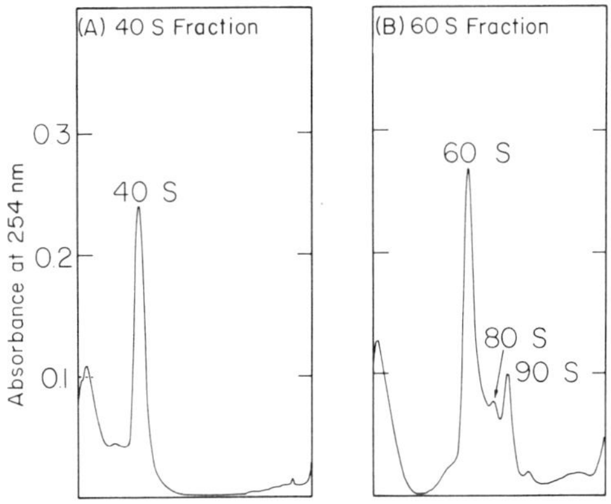

Fig. 3. Determination of the ribosomal particles present in the 40 and 60 S fractions from dissociated muscle ribosomes. The particles present in the 40 S (A) and 60 S (B) fractions isolated from a sucrose gradient similar to that in Figure 1A were determined by centrifugation of the fractions through sucrose gradients in medium A. The absorbance at the top of the gradient is attributable to β-mercaptoethanol.

subunits are prepared with puromycin, contamination of the 60 S fraction is less than 5%.

If ribosomal subunits are to be used shortly after isolation, they are dialyzed overnight at 4°C against 200 vol of medium A containing 20 mM β-mercaptoethanol. Reassociated ribosome monomers are formed by combining equimolar amounts of subunits before dialysis; the ratio (60 S/40 S) is approximately 2.5:1 (A_{260} units). Reassociation of muscle ribosomal subunits is nearly 100% efficient. Reassociated ribosomes can also be formed, albeit less efficiently, by mixing dialyzed subunits just prior to assay.

The ribosome subunits can be concentrated. The fractions are dialyzed briefly (1-2 hr) against medium A containing 20 mM β-mercaptoethanol; 0.2 vol of ethanol is added to precipitate the subunits, and the mixture is kept at 0°C for 1 hr. Dialysis is required to lower the concentrations of sucrose and potassium which interfere with precipitation of ribosomal particles by ethanol. The concentrations of magnesium and ethanol are critical. At the concentration of magnesium in medium A (12.5 mM), 0.2 ml of ethanol precipitates at least 95% of the subunits and there is no loss of activity; lower concentrations of ethanol decrease the yield of subunits; higher concentrations precipitate all the particles but their activity is diminished. At lower magnesium concentrations higher concentrations of ethanol are required to precipitate all the subunits; again, the activity of the particles is decreased. The precipitated ribosomes are collected by centrifugation at 8000 X g (in a Servall Hb-4 rotor at 7000 rpm) for 15 min at 4°C. After the supernatant is discarded and excess ethanol is wiped from the centrifuge tube, the pellet is resuspended in medium A containing 20 mM β-mercaptoethanol. An alternative means of concentrating subunits is by ultrafiltration.

Subunits in medium A containing 20 mM β-mercaptoethanol remain active for approximately 1 week when stored at 4°C, but suspensions of subunits lose activity if frozen and thawed.

Subunit pellets can be stored for long periods of time. The subunit fractions are diluted with 3 vol of buffer containing 50 mM tris-HCl (pH 7.5), 12.5 mM $MgCl_2$, and 20 mM β-mercaptoethanol and centrifuged at 105,000 X g (in a Spinco 40 rotor at 40,000 rpm) for 16 hr at 4°C. After centrifugation the tubes are drained, wiped dry, and the pellet is kept at -70°C; the subunits remain active for at least 6 weeks. Just prior to experiments the subunits are suspended in medium A containing 20 mM β-mercaptoethanol by gentle homogenization, incubated for 5 min at 37°C, and aggregates are removed by centrifugation at 3000 X g for 10 min.

A critical factor in the preparation of subunits from rat muscle ribosomes is the temperature of centrifugation. The sedimentation profile obtained at 28°C is shown in Fig. 1; at 4°C a large number of 90 and 105 S particles are formed.

(The 90 S particles are dimers of the 60 S subunit; the 105 S particle is probably an aggregate of a 40 S and two 60 S subunits.) Clearly, when gradients are run at 4°C, there is decreased resolution of the particles and a lower yield of 40 and 60 S subunits. Aggregation of ribosomal particles differs according to the species from which they are derived; rabbit muscle ribosomes do not aggregate at 4°C in sucrose gradients [19].

V. ASSAY OF PROTEIN SYNTHESIS BY MUSCLE RIBOSOMES

Maximal protein synthesis by skeletal muscle ribosomes requires aminoacyl-tRNA (or a system capable of generating aminoacyl-tRNA from free amino acids and tRNA), supernatant protein factors, GTP, an energy-generating system, ATP, magnesium, and an SH-protecting reagent [10]. Some of the components are not essential (as, for example, ATP and an energy-generating system); rather, they increase the amount of protein that is synthesized.

We measure protein synthesis by incubating ribosomes at 37°C in 50 mM tris-HCl (pH 7.8), 80 mM KCl, 5-20 mM $MgCl_2$, 10 mM β-mercaptoethanol, 5 mM ATP, 0.05 mM GTP, 1 mM phosphoenolpyruvate, 10 μg of pyruvate kinase, 100 μg of tRNA acylated with 1 radioactive amino acid and 19 nonradioactive amino acids, and 2 mg of muscle supernatant protein; the total volume is 1 ml. (The volume of the reaction mixture is not critical; the entire reaction can be performed in 0.05 or 0.1 ml if substrates must be conserved.) Protein synthesis is very rapid for the first 5 min of incubation; thereafter the rate declines, and after 20 min no further synthesis occurs.

We have always used 50 mM tris-HCl (pH 7.8) as a buffer, mainly as a matter of habit. The pH optimum for protein synthesis is broad, from 7 to 8. Any pH within that range is satisfactory. Actually, the tris concentration can be as low as 20 mM without affecting protein synthesis. Surely other buffers can be used.

A monovalent cation, either potassium or ammonium (they appear to be interchangable) is required for protein synthesis. The peptidyl transferase reaction requires a monovalent cation [20]; perhaps other steps in protein synthesis do also. Protein synthesis is maximal over a broad range of concentrations (50-150 mM) of potassium or ammonium. Sodium and lithium cannot replace potassium; in fact, they are inhibitory.

A divalent cation (magnesium is almost always used) is absolutely essential for protein synthesis by ribosomes. Magnesium is required for the structural integrity of ribosomes and for the binding of mRNA and aminoacyl-tRNA to ribosomes. The concentration of magnesium for optimum protein synthesis varies with the mRNA and the amount of ATP and GTP in the reaction mixture. (Nucleoside triphosphates chelate divalent cations and thus lower the effective concentration of magnesium.)

Optimum concentrations of magnesium with the templates we have used are discussed below. It is likely that soluble calcium salts can be used in place of magnesium; at least calcium can be used as the sole divalent cation when protein synthesis by bacterial ribosomes is assayed [21].

At least one of the enzymes required for the synthesis of protein, aminoacyl transferase II, is inactivated if its SH group(s) are oxidized [22]. SH groups in the ribosomal proteins may also need to be kept reduced [23]. For these reasons β-mercaptoethanol is in the reaction mixture.

GTP is required for a number of reactions: the initiation of protein synthesis, aminoacyl transferase I-catalyzed binding of aminoacyl-tRNA, and aminoacyl transferase II-catalyzed translocation of peptidyl-tRNA. The optimum concentration of GTP is 0.05 mM.

GTP is hydrolyzed during protein synthesis, therefore a means of regenerating the nucleoside triphosphate is needed; we use phosphoenolpyruvate and pyruvate kinase. Adamson *et al.* [24] found creatine phosphate and creatine phosphokinase to be superior for energy generation when protein synthesis by reticulocyte lysates is measured; we have not tested creatine phosphate and creatine phosphokinase with muscle ribosomes.

We do not know the role of ATP in the synthesis of protein from aminoacyl-tRNA by ribosomes; nonetheless, there is no doubt that ATP markedly increases the amount of protein synthesized. There is no evidence that ATP is needed for the initiation of peptide chains, for elongation of peptides, or for termination of the process. ATP may serve in the regeneration of aminoacyl-tRNA from tRNA deacylated during incubation (a process we know occurs), or it may participate in the regeneration of GTP from GDP.

Aminoacyl-tRNA is an obligatory intermediate in protein synthesis. Amino acids are activated and covalently linked to specific tRNA molecules. The aminoacyl-tRNA then recognizes the appropriate codon in mRNA for the amino acid it carries. We use commercial preparations of *Escherichia coli* B tRNA. The tRNA is aminoacylated with 1 radioactive amino acid and 19 nonradioactive amino acids; we use a crude preparation of *E. coli* B aminoacyl-tRNA synthetases [25]. The radioactive amino acid is usually phenylalanine (for no other reason than that it is the proper substrate for poly U-directed synthesis). In the reaction mixture given above, 100 μg of aminoacyl-tRNA is not saturating. However, protein synthesis is a linear function of ribosome concentration in the assay; thus the concentration of aminoacyl-tRNA, while not saturating, is not limiting either.

If an unresolved ribosome-free cytosol preparation is used, aminoacyl-tRNA can be generated from tRNA and amino acids by the action of aminoacyl-tRNA synthetases. The formation of aminoacyl-tRNA then occurs *pari passu* with protein synthesis

[26]. We have not used amino acids as substrates for protein synthesis by muscle ribosomes because it complicates the analysis of the mechanism of regulation by introducing a number of additional variables: the amount of endogenous tRNA and the amount and activity of the aminoacyl-tRNA synthetases.

Muscle supernatant contains enzymes required for protein biosynthesis: aminoacyl transferase I, which catalyzes binding of aminoacyl-tRNA to the mRNA-ribosome complex, and aminoacyl transferase II, which catalyzes translocation of peptidyl-tRNA from the acceptor to the donor site on the ribosome and results in movement of the ribosome along mRNA. To prepare muscle supernatant the initial homogenization of the tissue is performed in medium A containing 250 mM sucrose, and ribosomes are sedimented by centrifugation at 78,000 X g (see Section III). After a second centrifugation at 105,000 X g (in a Spinco 40 rotor at 40,000 rpm) for 2.5 hr at 4°C to remove residual ribosomal particles, the pH of the supernatant is adjusted to 5.2 (at 4°C) by dropwise addition of 1 N acetic acid. The precipitate (which contains tRNA, some of the aminoacyl-tRNA synthetases, and perhaps the remaining ribosomes) is removed by centrifugation at 13,000 X g (in a Servall GSA rotor at 9000 rpm) for 15 min at 4°C. The pH 5.2 supernatant is brought to pH 7.0 with 1 N KOH and stored in small aliquots at -20°C; the enzymes are stable for at least 6 months. Under the conditions we use to assay muscle ribosomes, 2 mg of supernatant protein give optimal protein synthesis; higher concentrations decrease the total amount of amino acid incorporated.

Protein synthesis is absolutely dependent on ribosomes—the enzyme that catalyzes peptide bond formation (peptidyl transferase) is a 60 S ribosomal protein [27]—and a template RNA [26]. The template has a series of codons which are translated into a sequence of amino acids. Protein synthesis can be directed by endogenous mRNA that remains on the muscle ribosomes during isolation. In all likelihood, when the template is endogenous mRNA, there occurs only elongation of nascent peptides whose synthesis has been initiated in the cell. Reinitiation of translation of endogenous mRNA is unlikely to occur _in vitro_. It is very convenient to keep the concentration of ribosomes limiting; 60 μg (0.67 A_{260} units) of ribosomes is a good amount to use to measure endogenous protein synthesis, but it is probably best to determine a ribosome concentration curve (concentration versus incorporation activity) before undertaking experiments. The amount of endogenous incorporation varies with the magnesium concentration (see above); optimum synthesis is at 10-12.5 mM (Figure 4).

Exogenous templates can also be used to direct protein synthesis by ribosomes; for example, in the presence of the synthetic polynucleotide polyuridylic acid (poly U), polyphenylalanine is synthesized. With the addition of 100 μg

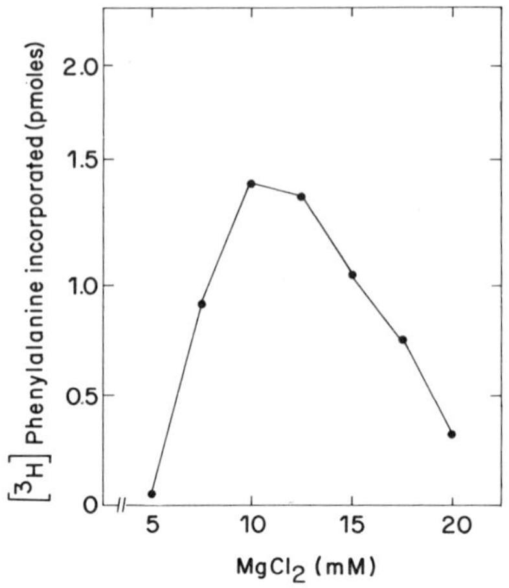

Fig. 4. Effect of the concentration of magnesium on endogenous protein synthesis. For the assay, 60 μg of ribosomes were used.

---

of poly U to the usual reaction mixture, the synthesis of polyphenylalanine is proportional to the amount of ribosomes, up to 20 μg of ribosomes (0.22 A_{260} units). Optimum synthesis is at a magnesium concentration of 15-17.5 mM (Figure 5. Other synthetic polynucleotides can probably serve as templates, but we have not tried them. Ribosomes can be made completely dependent upon exogenous template RNA by removing endogenous nascent peptide and, presumably, endogenous mRNA. This can be done by preincubating ribosomes (22 A_{260} units/ml) at 37°C for 30 min in medium A containing ATP, GTP, phosphoenolpyruvate, pyruvate kinase, and β-mercaptoethanol (in the same concentrations as for protein synthesis), together

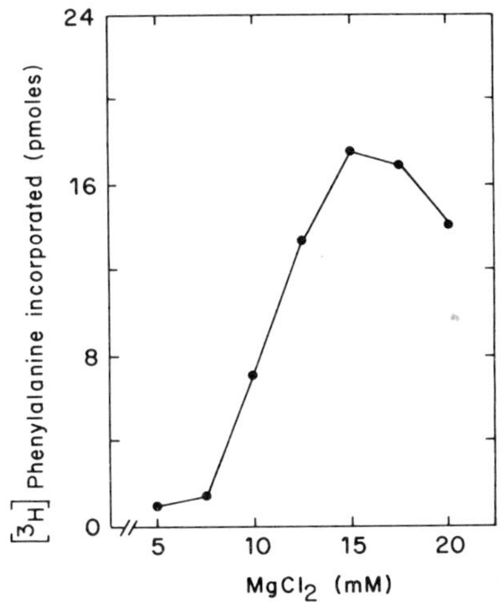

Fig. 5. Effect of the concentration of magnesium on poly U-directed polyphenylalanine synthesis. For the assay 15 μg of ribosomes and 100 μg of poly U were used. The amount of endogenous incorporation has been subtracted.

---

with supernatant protein (4 mg/ml) and puromycin (50 μg/ml). Ribosomes are recovered by layering the mixture onto 5 ml of 500 mM sucrose in medium A and centrifuging at 150,000 X g (in a Spinco Ti-50 rotor at 50,000 rpm) for 3 hr at 4°C. An alternative procedure is to use ribosomes reassociated from subunits.

When ribosomes that have been reassociated from isolated subunits are used to assay protein synthesis, an exogenous template is obligatory (Table I). Apparently, dissociation leads to loss of endogenous mRNA. Of course, the individual subunit fractions are not capable of peptide synthesis; both subunits are necessary. [The small amount of polyphenylalanine synthesized by the 60 S subunit fraction (Table I) results from contamination with 40 S subunits.] The optimum magnesium concentration for the translation of poly U by reassociated ribosomes is 15-17.5 mM. Reassociated ribosomes synthesize very large amounts of polyphenylalanine, presumably because they lack

TABLE 1

Protein Synthesis by Subunits from Rat Muscle Ribosomes[a]

	Protein synthesis (pmoles [^{3}H]phenylalanine incorporated per 20 μg ribosomes)	
Ribosomal particle	Endogenous synthesis	Polyphenylalanine synthesis
Untreated ribosomes	1.21	10.78
40 S Fraction	0	0.41
60 S Fraction	0.04	2.52
40 + 60 S Fractions	0.04	12.45

[a]For the assay 15-25 μg of ribosomal particles were used; the concentration of $MgCl_2$ was 15 mM; the amount of poly U when present was 100 μg. The amount of phenylalanine incorporated in the absence of poly U ("blank" incorporation) was taken into account in calculating polyphenylalanine synthesis.

endogenous mRNA and are thus able to bind and translate more poly U.

After incubation of the ribosomes, protein synthesis is stopped by addition of 3 ml of 10% trichloroacetic acid (TCA). The samples are heated at 90°C for 15 min to hydrolyze tRNA (including aminoacyl- and peptidyl-tRNA); the only remaining materials insoluble in TCA are peptides. (An alternative means of hydrolyzing tRNA is to add 0.1 vol of 3 N KOH to the reaction mixture and to incubate it for 30 min at 37°C; 0.1 vol of 3 N HCl is then added to neutralize the KOH and 2 ml of 10% TCA are used to precipitate peptides.) After the sample is chilled at 0°C for 10 min, protein is collected on glass fiber discs (Whatman GF/C filters, 24 mm diameter) held in Millipore filtration units. Glass-fiber discs are as efficient as nitrocellulose filters (which are often used) for collecting protein precipitates and are substantially less expensive. The protein precipitate collected on the filter is washed with four 10-ml portions of 5% TCA containing 10 mg/ml of nonradioactive amino acid corresponding to the radioactive amino acid used in the assay. The washing and the presence of the nonradioactive amino acid decrease the amount of radioactive amino acid that might be nonspecifically adsorbed to the filter. The filter discs are than put into glass vials containing 0.7 ml of 90% formic acid; the acid solubilizes the protein on the filters. (We have tried to measure the radioactivity on the filter in

Bray's scintillation fluid without solubilizing the protein but have had no success; the difficulty is that the radioactivity slowly increases with time, presumably because of slow solubilization of some of the precipitated protein.) After the protein is dissolved, 15 ml of scintillation fluid (0.5% 2,5-bis[2-(5-tert-butylbenzoxazolyl)]thiophene (BBOT) in 50:50 toluene-methyl Cellosolve) is added. Radioactivity is measured in a liquid scintillation spectrometer.

Precipitation with TCA limits somewhat the measurement of peptides synthesized by ribosomes. For example, polymers of phenylalanine containing six residues or more are quantitatively precipitated by 10% TCA; however, short oligophenylalanyl peptides (two to five residues) are soluble in 10% TCA.

REFERENCES

[1] A. Korner, Nature, 181, 422 (1958).

[2] I. G. Wool, W. S. Stirewalt, K. Kurihara, R. B. Low, P. Bailey, and D. Oyer, Recent Prog. Hormone Res., 24, (1968).

[3] T. E. Martin, F. S. Rolleston, R. B. Low, and I. G. Wool, J. Mol. Biol., 43, 135 (1969).

[4] J. R. Florini and C. B. Breuer, Biochemistry, 5, 1870 (1966).

[5] W. S. Stirewalt, I. G. Wool, and P. Cavicchi, Proc. Nat. Acad. Sci. U.S., 57, 1885 (1967).

[6] A. G. Szent-Györgyi, in Structure and Function of Muscle (G. H. Bourne, ed.), Academic Press, New York, 1960, p. 1.

[7] S. M. Heywood, R. M. Dowben, and A. Rich, Proc. Nat. Acad. Sci. U.S., 57, 1002 (1967).

[8] R. M. Dowben, T. A. Gaffey, and P. M. Lynch, FEBS Letters, 2, 1 (1968).

[9] G. Bullock, A. M. White, and J. Worthington, Biochem. J., 108, 417 (1968).

[10] I. G. Wool and P. Cavicchi, Biochemistry, 6, 1231 (1967).

[11] A. Fleck and H. N. Munro, Biochim. Biophys. Acta, 55, 571 (1962).

[12] T. E. Martin and I. G. Wool, Proc. Nat. Acad. Sci. U.S., 60, 569 (1968).

[13] H. Lamfrom and E. Glowacki, J. Mol. Biol., 5, 97 (1962).

[14] Y. Tashiro and T. Morimoto, Biochim. Biophys. Acta, 123, 523 (1966).

[15] A. K. Falvey and T. Staehlin, J. Mol. Biol., 53, 1 (1970).

[16] W. S. Stirewalt, J. J. Castles, and I. G. Wool, Biochemistry, in press.

[17] T. E. Martin, I. G. Wool, and J. J. Castles in Methods in Enzymology, Vol. 12, Part C (L. Grossman and K. Moldave, eds.), Academic Press, New York, 1971.

[18] W. Gilbert, J. Mol. Biol., 6, 389 (1963).

[19] T. E. Martin and I. G. Wool, J. Mol. Biol., 43, 151 (1969).

[20] B. E. H. Maden and R. E. Monro, European J. Biochem., 6, 309 (1968).

[21] J. Gordon and F. Lipmann, J. Mol. Biol., 23, 23 (1967).

[22] R. D. Mosteller, J. M. Ravel, and B. Hardesty, Biochem. Biophys. Res. Commun., 24, 714 (1966).

[23] H. C. McAllister and R. S. Schweet, J. Mol. Biol., 34, 519 (1968).

[24] S. D. Adamson, E. Herbert, and W. Godchaux, Arch. Biochem. Biophys., 125, 671 (1968).

[25] G. von Ehrenstein and F. Lipmann, Proc. Nat. Acad. Sci. U.S., 47, 941 (1961).

[26] M. Nirenberg and J. H. Matthaei, Proc. Nat. Acad. Sci. U.S., 47, 1588 (1961).

[27] D. Vazquez, E. Battaner, R. Neth, G. Heller, and R. E. Monro, Cold Spring Harbor Symp. Quant. Biol., 34, 369 (1969).

Chapter 2

ISOLATION OF MAMMALIAN CELL POLYRIBOSOMES

Ralph B. Arlinghaus and Richard Ascione

Department of Biology
The University of Texas
M. D. Anderson Hospital and Tumor Institute
Houston, Texas
and
Plum Island Animal Disease Laboratory
Animal Disease and Parasite Research Division
Agricultural Research Service
U. S. Department of Agriculture
Greenport, New York

I. INTRODUCTION 30

A. General Statements on Ribosome Preparation . . 31

B. Cell Lysis 32

C. Ionic Environment. 33

D. Technique Used to Inhibit RNase. 34

II. ISOLATION OF RABBIT RETICULOCYTE POLYRIBOSOMES . . 35

III. PREPARATION OF RIBOSOMES FROM RETICULOCYTES FREE OF TRANSFER FACTORS. 37

A. Salt Shock and Preincubation 37

B. DOC Treatment. 39

C. Properties 40

IV. PREPARATION OF POLYRIBOSOMES FROM A CULTURED MAMMALIAN CELL 40

A. Experimental Procedure 42

B. Tumor Cell Polyribosomes 44

V. PREPARATION OF HEPATIC POLYRIBOSOMES 44
A. Liver Isolation and Preparation. 46
B. Cell Lysis and Isolation of Ribosomes. . . . 47
C. Ribosome and Polysome Purification 50
VI. AMINO ACID INCORPORATION SYSTEM. 51
VII. PREPARATION OF ENZYMES FOR PROTEIN SYNTHESIS . . 52
VIII. SUCROSE DENSITY GRADIENT CENTRIFUGATION. 53
REFERENCES 55

I. INTRODUCTION

Ribosomes are complex macromolecular structures that play a major role in the all-important task of translating genetic messages into protein. Ribosomes of bacterial and animal cells have been widely studied, and many reviews on their structure and function have been compiled [1—3]. This chapter discusses methods of preparing mammalian cell polyribosomes from several sources, specifically, rabbit reticulocytes, liver, and baby hamster kidney (BHK) cells grown in tissue culture. In addition, a method of washing ribosomes free of transfer factors is presented, and the preparation of soluble enzymes that function in protein synthesis is also described.

Mammalian cell ribosomes contain two subunits, often denoted or identified by their approximate sedimentation rate in dilute salt solutions as 60 and 40 S. The larger subunit contains one equivalent of the 28 S rRNA, and the smaller particle contains one equivalent of the 18 S rRNA. In

Escherichia coli, the structure of ribosomal subparticles (50 and 30 S) is being actively investigated, and some preliminary reports indicate that the smaller subunit contains 20 discrete proteins and that the larger subunit may contain about 36 proteins [4]. For methods of preparation of ribosomal subparticles, see Chapter 1, Section IV.

A. General Statements on Ribosome Preparation

In our experience, several factors must be considered before deciding on the method of ribosome isolation. First, for what use are the ribosomes intended? Second, do the cells from which the ribosomes are to be extracted contain nuclei? Third, do the cells contain mainly membrane-bound ribosomes? Last, are there high levels of RNases present in the cell used for ribosome preparation? Each of these factors is important and must be considered in the selection of an isolation procedure.

Enucleated cells, exemplified by rabbit reticulocytes, require very different methods (simpler) than nucleated cells, such as liver cells or kidney cells. The state of ribosomes within the cell is also a major consideration. In liver cells most of the ribosomes are membrane bound. In rabbit reticulocytes the majority of ribosomes are membrane-free. Ribosomes from cells grown in culture might be similar to rabbit reticulocytes (i.e., membrane-free) or liver cells (membrane bound), but usually both types are represented. For example, HeLa cells contain 85% free ribosomes; 15% are membrane bound [5]. Thus the methods of cell lysis and subsequent differential centrifugation are vastly different for nucleated cells with membrane-bound ribosomes than for the same type of cell with membrane-free ribosomes. (See Chapter 3, Section VIII, for details on methods of preparation of free and membrane bound ribosomes and polyribosomes.)

Polyribosomes have the greatest cell-free activity when they are isolated by gentle procedures in a crude form. If the native activity of a preparation is of interest, these crude polyribosomes are perhaps the type of preparation needed. If one is interested in adding mRNA to ribosomes with the goal of synthesizing finished protein coded by this RNA message, crude polyribosomes suffice only if a means of identifying the desired labeled protein in a mixture of other labeled proteins is available [6]. Otherwise it is necessary to use ribosome preparations devoid of native activity, which have the appropriate initiator proteins [7—9] to permit translation of the desired message. Purified, and usually less active, ribosomes are required in studies concerned with transfer factors or initiating factors, or the identification of ribosomal proteins.

B. Cell Lysis

The method of cell breakage and homogenization must be carefully considered. From a consideration of polyribosome morphology, it is apparent that these structures are quite labile to mechanical fragmentation; therefore, high shearing forces are to be rigorously avoided. Generally, this means avoidance of freezing, sonication, blending (e.g., Waring Blendor), or the use of pressure cells to disrupt mammalian cells. In general, techniques involving hypotonic lysis or nonionic detergents, or even mild homogenization, yield the best polyribosomal profiles.

C. Ionic Environment

The ionic environment during polyribosomal isolation and analysis is quite critical, specifically with respect to monovalent and divalent cations. Optimal conditions appear to vary with different mammalian preparations [10]. The optimal ratio of monovalent/divalent cation ranges from 10:1 to 20:1 and, in general, the optimal divalent cation (magnesium ion) concentration ranges from 3 to 10 mM. For analytical work, however, 10 mM Mg^{2+} is rather high and tends to produce aggregates that are deceptively similar to polyribosomes. Bacterial ribosomes, unlike mammalian ribosomes, require Mg^{2+} concentration of 10 mM or higher for stability. As a rule, we have found that including EDTA in the magnesium buffer at 5—10% of the concentration of Mg^{2+} serves to chelate any undesirable divalent ions present in the buffers; Zn^{2+}, Cu^{2+}, and Pb^{2+} are quite detrimental to polyribosome stability because (nonenzymic) phosphodiester cleavage is catalyzed by these divalent cations.

Caution should also be used in employing certain buffers, especially ones with the ability to chelate divalent ions (e.g., tris—maleate, tris—succinate, phosphate, or pyrophosphate buffers). In fact, even high concentrations of anionic detergents such as deoxycholate (DOC), dextran sulfate, or Duponol (sodium dodecyl sulfate or SDS) require an increase in the concentration of divalent cations to avoid polyribosome destabilization as a result of magnesium chelation. When possible, such buffers and detergents must be used judiciously.

D. Technique Used to Inhibit RNase

The bane of all those who wish to prepare polyribosomes is their extreme sensitivity to RNase. At trace concentration of RNase ($\sim$0.01 μg/ml) polyribosomes can be converted to monomeric units. Therefore: all operations should take place at cold temperatures; the release of RNases by rupturing lysosomes and membranes should be avoided if possible; and subcellular components should be fractionated as quickly as possible to minimize nuclease degradation of polyribosomes. In the case of attachment or interaction of polyribosomes with membranes, a further complication arises. For hepatic cells, with a well-developed endoplasmic reticulum, detergents must be employed in order to increase substantially the yields of ribosomes. Several mild detergents are available which successfully release membrane bound ribosomes, for example, sodium DOC,Triton X-100 (Rohm and Haas), NP-40 (Shell Chemical Company, New York, New York), Lubrol WX (A. H. Hoffman, Inc., Providence, Rhode Island), or combinations of these. Although such detergents are mild enough to yield stable polyribosomal structures, by solubilizing membranes, they release a host of lytic enzymes (particularly a membrane-associated RNase) which can, and often do, degrade polyribosomes. Degradation can be avoided by several means, for example, by the judicious use of RNase inhibitors such as synthetic polyanions (e.g., polyvinyl sulfate or dextran sulfate) or natural inhibitors (e.g., ribosome-free hepatic cell sap or heparin [11]).

II. ISOLATION OF RABBIT RETICULOCYTE POLYRIBOSOMES

Rabbits are made anemic with phenylhydrazine as previously described [12]. In our laboratory best results are obtained by subcutaneous injection on each of 4 days of 1.0 ml of 2.5% phenylhydrazine per 5—6 lb New Zealand white rabbit, followed by no injections on the fifth and sixth days, and killing of the rabbits on the seventh day (see also Chapters 8 and 9). One milliliter of heparin (200 units/ml, Mann Biochemicals) is injected into the chest cavity prior to heart puncture, and the syringe and collecting vessels are treated with heparin to prevent blood clotting. The phenylhydrazine solution is neutralized to pH 7.0 with NaOH, adjusted to 1 mmole/liter in glutathione, and stored frozen at -20°C. Each day's supply is stored in a separate vial to eliminate the necessity for thawing and refreezing. Blood is removed by heart puncture, and the cells are pelleted by centrifugation at 2000 X g in a GSA rotor for 20 min at 4°C. All subsequent steps are carried out at

0—4°C. The cells are resuspended in a volume of NKM (0.13 M NaCl, 5 mmoles/liter KCl, 7.5 mmoles/liter $MgCl_2$) solution equal to the amount of plasma decanted, and the cells are again isolated by centrifugation. The reticulocytes are lysed by osmotic shock in dilute Mg^{2+}. The time period was originally set for 10 min in the dilute Mg^{2+} solution [12], but Marks et al. [13] found that reducing the osmotic shock to 1.5—2.0 min decreased the lysis of cells other than the reticulocytes, particularly white cells. For lysis the packed cells are suspended in 4 vol of 2.5 mmoles/liter $MgCl_2$ and stirred vigorously for 1.5 min. One volume (equal to the volume of packed cells) of a solution containing 1.5 M sucrose and 0.15 M KCl is added and the suspension centrifuged at 2500 X g for 20 min. The addition of KCl—sucrose halts the osmotic shock, keeping most of the white cells intact and permitting them to be removed from the solution by centrifugation. The ribosomes are then isolated from this supernatant by sedimentation at 78,000 X g for 90 min. Generally, 70—80 ml of packed cells are isolated from six rabbits to yield about 200—250 mg of crude ribosomes. A solution containing 1 mg/ml of ribosomes has an absorbance of 12.0 at 260 nm. The ribosome pellet is rinsed twice with 0.25 M sucrose and suspended by gentle homogenization in the same solution. The preparation is clarified prior to storage at -20°C by an 8000 X g spin for 5 min. Preparations are stable at -20°C for at least several months. These preparations are composed of 65—75% polyribosomes, 20—25% ribosomes, and the balance consists of ribosome subparticles. A representative sucrose gradient profile is shown in Figure 1B. The *in vitro* activity of these preparations ranges from 5 to 16 nmoles of total amino acid per milligram of ribosomes. To obtain more active preparations for amino acid incorporation, the reticulocytes can be lysed in 1—2 vol of a 2.5 mmoles/liter Mg^{2+} buffer and then adjusted to isotonicity; the intact cells are pelleted by centrifugation as above. This lysate [6] has been used successfully by many investigators, but the results in our laboratory have been somewhat variable for amino acid incorporation studies.

III. PREPARATION OF RIBOSOMES FROM RETICULOCYTES FREE OF TRANSFER FACTORS

A. Salt Shock and Preincubation

This procedure includes incubation in KCl (salt shock), incubation in the complete amino acid-incorporating system (Ref. 12 and Section VI), and treatment with DOC. The ribosomes are pelleted once after the salt shock and incubation, and once again after the DOC treatment. This procedure inactivates the globin mRNA and removes activating enzymes, transfer factors, and probably initiator proteins [9].

Once-pelleted (1X) rabbit reticulocyte ribosomes, active in hemoglobin synthesis, are prepared as described. These 1X

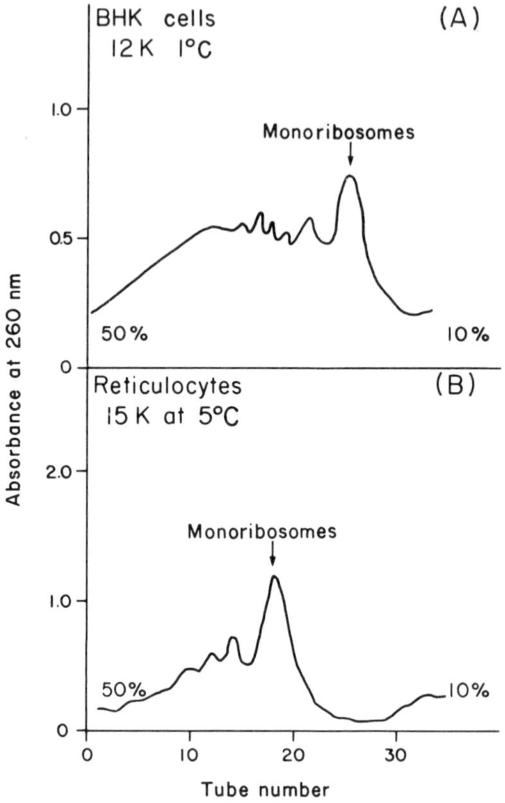

Fig. 1. Sucrose gradient profiles of rabbit reticulocyte and BHK polyribosomes. (A) Twice-pelleted polyribosomes (2X) from BHK cells were extracted and purified as described in text. They were analyzed on a 10—50%, linear, sucrose gradient buffered with 0.01 M tris (pH 7.5) which contained 1 mmole/liter $MgCl_2$, 10 mmoles/liter KCl, and 0.1 mmole/liter EDTA. The direction of sedimentation is from right to left, and the samples were centrifuged in a Spinco SW-25.1 rotor at the speed in rpm and temperature shown. (B) Once-pelleted (1X) rabbit reticulocyte polyribosomes were analyzed as in (A).

ribosomes are then incubated for 20 min at 37°C in a mixture containing: 46 mM tris--Cl (pH 7.5), 92 mM KCl, 18 mM GSH, and 8.7 mg/ml 1X ribosomes. After incubation the flask is chilled, and additional components are added to give the following mixture: 37 mM tris--Cl (pH 7.5), 74 mM KCl, 15 mM GSH, 7.0 mg/ml 1X ribosomes, 3.7 mM $MgCl_2$, 0.74 mM ATP, 5.5 mM creatine phosphate, 22 μg/ml creatine kinase, 37 μM each of all 20 common amino acids, and 9 mg/ml AS_{70} enzyme (see Section VII or Ref. 12). This reconstitutes the complete amino acid incorporation system. The mixture is incubated 40 min at 37°C, chilled, and diluted with approximately 6.3 vol of a medium containing 0.25 M sucrose, 8 mM $MgCl_2$, 70 mM $KHCO_3$, and 0.1 M KCl. The ribosomes are then collected by centrifugation at 78,000 X g for 90 min. The pelleted ribosomes are carefully homogenized, and aggregates removed by centrifugation at 10,000 X g for 10 min. The ribosome concentration is adjusted to 14 mg/ml, using a specific extinction coefficient of 11.2 absorbance units/mg at 260 nm.

B. DOC Treatment

DOC treatment is necessary for the complete removal of transfer enzymes, particularly TF-2, from the ribosomes. The shocked and preincubated ribosomes are adjusted to 1% with sodium DOC (Mann Biochemicals; 0.25 vol of a 5% stock solution) and incubated at 37°C for 3 min. The mixture is chilled and diluted 10-fold with a medium containing 8 mM $MgCl_2$, 70 mM $KHCO_3$, and 0.25 M sucrose. The ribosomes are then isolated by centrifugation at 78,000 X g for 90 min. The pelleted ribosomes are homogenized in 0.25 M sucrose and centrifuged at 10,000 X g for 10 min to remove aggregates. The concentration is adjusted to 15 mg/ml, and the ribosomes are stored at -90°C. This preparation is termed 3X DOC ribosomes.

C. Properties

The 3X DOC ribosomes are essentially free of activity for globin mRNA and transfer enzymes. Omission of poly U, the binding enzyme (TF-1), or TF-2 from any of the assays described for the transfer enzymes [14] results in less than 10% of the activity (binding or peptide formation) observed in the presence of these components. The ribosomes seem to be active with various mRNA species; for example, in the presence of poly U, poly A, and poly C, the ribosomes bind phenylalanyl-, lysyl-, and prolyl-tRNA, respectively, but peptide bond formation has been shown only with poly U. Peptide bond formation with these ribosomes is susceptible to inhibition by puromycin and cycloheximide, but not by chloramphenicol. The assay for this preparation has been described previously [14].

IV. PREPARATION OF POLYRIBOSOMES FROM A CULTURED MAMMALIAN CELL

This procedure [7] was originally designed for the isolation of polyribosomes from cells for use in amino acid incorporation studies and has been used successfully with several different cell lines, both primary and continuous, such as primary bovine kidney, pig kidney (PK), JLS-V5 (a Rauscher leukemia virus-transformed cell from spleen-thymus), and BHK. However, most of the methodology was established with BHK-21 clone 13 [15] cells and the discussion that follows applies only to BHK cells. (See Chapter 3 for methods used for these other cell lines.) Two types of problems are encountered in the isolation procedure. First, a large proportion of ribosomes are apparently membrane bound and, second, nuclease levels in the cell preparations are high. These two problems are laborious and time-consuming to solve, since detergents must be used to free the ribosomes from cytoplasmic membranes; more RNase is thereby released, which degrades the polyribosomes and thereby compounds the isolation problem.

The procedure for polyribosome preparation involves three main factors that require consideration. First, a soluble RNase inhibitor (dextran sulfate 500, Pharmacia Fine Products, Uppsala, Sweden) is employed at low concentrations to inhibit RNase. Second, a hypertonic salt solution is used during cell lysis to help preserve the nuclear membrane; and third, Triton X-100 (or Lubrol or NP-40), in the presence of a low concentration of DOC is used to solubilize or free the polyribosomes from the large membrane fraction of the cell. Time is also a significant factor during the procedure, as is temperature. All operations must take place quickly at 0—4°C. The cytoplasm may or may not be treated with sodium DOC to remove any residual membranes bound to the polyribosomes. If DOC treatment is indicated, it is important to add additional Mg^{2+} and KCl to stabilize the polyribosomes prior to this treatment.

A. Experimental Procedure

Approximately 3—5 X 10^8 BHK cells (about 1 g wet weight) are rapidly cooled and scraped from tissue culture bottles by means of a rubber policeman. The cells are washed free of growth medium by resuspension in 200 ml of ice cold, buffered, bovine serum albumin solution containing 20 mM tris—HCl, 0.13 M KCl, 5 mM $MgCl_2$, and 5 mg/ml of bovine serum albumin, fraction V (adjusted to pH 7.3). The cells are resuspended at 0°C in 5 ml of hypotonic buffer solution containing 20 mM tris (pH 7.5), 50 mM NaCl, 3.6 mM $CaCl_2$, and 5 mM $MgCl_2$, and allowed to swell for 3 min. Then, an equal volume of 1.0% NP-40 (Shell Chemical Company) or Triton X-100 (Supplier: Sigma Chemical Company, St. Louis, Missouri) or Lubrol WX solution containing 40 mM tris, 0.50 M sucrose, 10 mM $MgCl_2$, 3.6 mM $CaCl_2$, 0.25 M KCl, and 1 mM EDTA (adjusted to pH 7.5) is added. The cellular slurry is mixed with a vortex mixer and lysed with three strokes of a loose-fitting Dounce homogenizer. Immediately after homogenization an RNase inhibitor such as dextran sulfate 500 is added to the lysate at a final concentration of 50 μg/ml, and the nuclear fraction is removed by centrifugation at 1000 X g for 5 min. To increase the yield of polyribosomes, the nuclei are washed again by Dounce homogenization of the 1000 χ g pellet in 10—20 ml of a solution containing 0.275 M sucrose, 0.10 M KCl, 50 mM NaCl, 3.6 mM $CaCl_2$, 5 mM $MgCl_2$, 0.5% Triton X-100 or Lubrol WX, 0.05% sodium DOC, 40 mM tris—HCl buffered at pH 7.5, and dextran sulfate (50 μg/ml). The washed nuclei are centrifuged as before, and the cytoplasmic supernatants combined. Cytoplasmic membranes are dispersed using 0.25—0.5% sodium DOC after adjusting the cation concentrations to 10 mM $MgCl_2$, 0.15 M KCl, and 5 mM $CaCl_2$ to stabilize the polyribosomes. The detergent-treated cytoplasm is centrifuged in a Spinco ultracentrifuge at 105,000 X g for 90 min, and a translucent polyribosome pellet is obtained. The surface of the polyribosome pellet (designated 1χ) is gently rinsed with 5 ml of a solution of 10 mM tris, 10 mM KCl, 1 mM $MgCl_2$, and 0.1 mM EDTA (pH 7.5) (tris—KCl—$MgCl_2$—EDTA buffer). For cell-free experiments, the polyribosomes are gently resuspended in 1—2 ml of the same buffer, using a Teflon pestle. The suspension is then layered over a discontinuous sucrose gradient (similarly buffered) composed of 10 ml of 0.5 M sucrose, 15 ml of 1.1 M sucrose, and 5 ml of 2.0 M sucrose. The polyribosomes are pelleted by overnight centrifugation at 25,000 rpm in a SW-25.1 rotor at 1°C and designated 2χ polyribosomes. (See Section V,C for a discussion of this step.)

These ribosomes are stored in NKM buffer at -20°C and are stable for long periods. They incorporate about 1—3 nmoles of total amino acids per milligram of ribosomes; a representative profile is shown in Figure 1A.

B. Tumor Cell Polyribosomes

The procedures described above for BHK polyribosomes can also be used with the JLS-V5 cell line transformed by Rauscher leukemia virus. A representative sucrose gradient profile of these polyribosomes is shown in Figure 2B.

V. PREPARATION OF HEPATIC POLYRIBOSOMES

We have modified the procedure described for BHK polyribosome isolation for use with various hepatic tissues, for example, rat liver, mouse liver, and guinea pig liver cells.

One factor markedly influences the distribution and isolation of ribosomes and polyribosomes from liver cells. Because of the relatively high content of glycogen stored in livers, most workers prefer to fast their animals for at least 12 hr prior to liver excision to deplete the polysaccharide content; this enhances the purity of the hepatic polyribosomes obtained. The isolation of liver polyribosomes is very sensitive to the nutritional environment within the cell, particularly with respect to the concentrations of amino acids [16]. In particular, hepatic polyribosomes are very sensitive to the concentration of tryptophan, which possibly plays an essential role in the regulation of liver protein synthesis, at least in so far as it affects the patterns of liver polysomes [17]. Consequently,

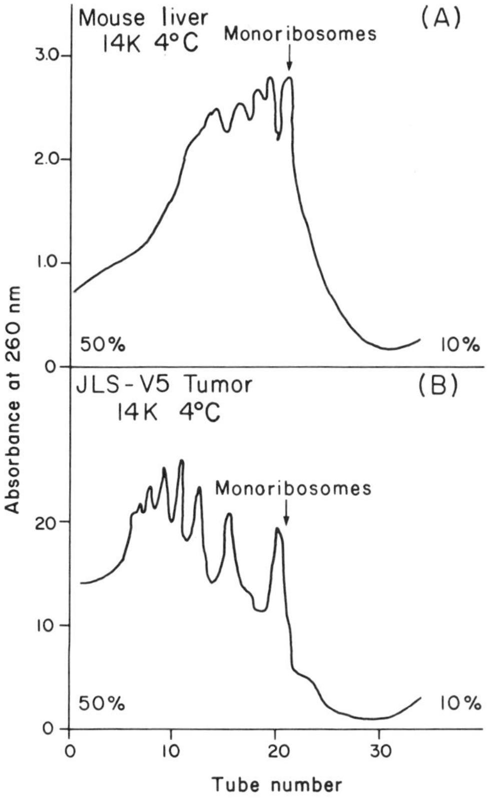

Fig. 2. Sucrose gradient profiles of polyribosomes from mouse liver and JLS-V5 tumor cells. (A) Twice-pelleted (2X) polyribosomes from mouse liver, obtained as described in the text, were analyzed as in Figure 1. (B) Twice-pelleted (2X) polyribosomes, obtained from Rauscher leukemia virus-transformed cells (JLS-V5), were analyzed as in Figure 1 but the sucrose gradient contained 10 mmoles/liter tris (pH 7.5) and 1 mmole/liter $MgCl_2$.

prolonged starvation results in alterations and loss of polyribosomes. There are two classes of polyribosomes, one more labile to starvation than the other [18]. Since the more stable class of polysomes is related to albumin synthesis, it appears that the class of polysomes labile toward starvation includes the species not bound to membranes (free). Therefore the effect of starvation illustrates one of several possible differential effects in liver polysome preparation that must be considered.

A. Liver Isolation and Preparation

In order to isolate liver cells, it is of course necessary to excise livers from the animals. Sacrifice and excision should be speedily accomplished, preferably in the absence of a lipophilic anesthetic (chloroform or ether) and, if possible, at or near 0°C. Use of a small animal guillotine (Harvard Apparatus Company), or simple manual decapitation, is deemed advantageous and, in some cases perfusion of the liver with extraction buffer is also recommended, although it is not absolutely necessary.

Immediately after the entire liver is excised, it should be blotted dry and trimmed (in the cold) free of any connective tissue, clotted blood, and gall bladder (except for rat livers) without releasing the biliary contents. The trimmed liver lobes should be weighed and finely minced using sterile stainless-steel surgical scissors. [From previous experience, we have found that boiling, autoclaving, or dry-heat sterilizing all materials that come in contact with cellular homogenates (e.g., glassware, homogenizers, dialysis tubing) reduces the chance of contamination by nucleases. RNase in particular should be regarded as ubiquitous, since humans possess an epidermal "finger" nuclease that can be solubilized by aqueous solutions or even adhere to glassware, thereby eventually introducing some RNA degradation. Consequently, the use of gloves when possible, especially in handling dialysis tubing or other objects coming in direct contact with polyribosomes, is strongly recommended.] The yield of minced liver ranges from approximately 2 g per adult mouse to 20 g per adult guinea pig, with that for the rat somewhere in between (~10 g per rat).

B. Cell Lysis and Isolation of Ribosomes

Suspend the minced liver in 2.5--3.0 vol of extraction buffer containing 0.30 M sucrose (RNase-free), 20 mM tris buffer (pH 7.3--7.5 at 25°C) or equivalent (for protein synthesis studies HEPES and TES buffers are deemed superior to tris [19]), 100 mM KCl, 3.5 mM $MgCl_2$, 3.5 mM $CaCl_2$, 0.7 mM EDTA, 0.5% Triton X-100 or a similar nonionic detergent, and 5 mM glutathione. The relatively high ionic strength helps to stabilize the polyribosomes for isolation and aids the Triton X-100 in detaching polysomes from membranes. The 0.30 M sucrose and 3.5 mM Ca^{2+} are present to preserve the integrity of the nuclear membranes during homogenization. Normally, stabilization against nuclear disruption is an essential factor in isolating ribosomes, since free ribosomes are able to bind rather nonspecifically to the chromosomal material released whenever nuclei lyse. Glutathione is utilized to stabilize the natural inhibitor of RNase present in the cytoplasm of most mammalian livers [11]. These inhibitors are readily inactivated by any reagent (particularly heavy metals) reacting with their free sulfhydryl groups.

The minced slurry is homogenized in 50-ml portions using a loose-fitting hand-operated Teflon or glass homogenizer similar to the one originally described by Dounce. Usually, 5 to 10 strokes are all that is necessary to disrupt the cells, provided that the original mince was finely divided. Improper or hasty mincing makes homogenization difficult, and care must be exercised so as not to break the glass homogenizer when such an instrument is used. After homogenization the mixture is made 50 μg/ml with respect to dextran sulfate 500, poured through a double layer of cheesecloth, and then centrifuged at 25,000 g-min (5000 × g for 5 min) to remove (sediment) the nuclei and cell debris. Although the natural hepatic nuclease inhibitor is present, the use of detergents solubilizes the particulate liver RNase necessitating additional protection in the form of the dextran sulfate. Haschemeyer and Gross [20] have reported that addition of low-molecular-weight yeast RNA (2—3 mg/ml) also helps to stabilize liver polyribosomes in the postnuclear supernatant when anionic detergent is used. The stabilizing effect of yeast RNA has been verified in this system, although it appears that dextran sulfate is a more potent inhibitor of nuclease.

If desired, the nuclear pellet may be resuspended and homogenized again in extraction buffer, using half the original volume. This procedure releases any additional polyribosomes associated with the sedimented nuclei. The postnuclear supernatant(s), containing polyribosomes at about 5 mg/g of original mince, is treated with sodium DOC to completely solubilize the polyribosomes from any membranes present. The postnuclear supernatant is adjusted to a final concentration of 1% in nonionic detergent,Triton X-100 or Lubrol WX, and 0.5% sodium DOC. All aqueous stock solutions of detergents (20% Lubrol WX and 10% sodium DOC) are stored frozen until use, and all buffers and solutions are stored in the cold after membrane filtration to eliminate the possibility of fungal or bacterial contamination.

Nonionic detergent treatment does not yield as active a cell-free preparation as the combination of both anionic and nonionic detergents. Therefore the combination of both types of detergent is recommended. To sediment ribosomes and polyribosomes from the postnuclear supernatant, an equal volume of extraction buffer, minus sucrose and Triton X-100 (or Lubrol WX), is added and the diluted supernatant is centrifuged (at 1°C) for about 10^7 g-min (100,000 X g average for 100 min). The pellet obtained is light to red brown in color and is translucent. The postribosomal supernatant is drained off and discarded and the tubes are wiped clean with tissue paper. The pellet should be rinsed once with several milliliters of TKMV buffer consisting of 10 mM tris (pH 7.5) (or equivalent good buffer, i.e., HEPES or TES), 10 mM KCl, 1.2 mM $MgCl_2$, and 0.1 mM EDTA. The yield of ribosomes is about 5 mg/g of original liver. The 1X ribosome pellets are gently resuspended in 0.3—0.5 ml TKMV buffer, for each gram of liver mince used, with a hand-operated Teflon homogenizer (Potter-Elvehjam type).

C. Ribosome and Polysome Purification

The resuspended polyribosomes are clarified by low-speed centrifugation to remove insoluble aggregates at 50,000 g-min (10,000 X g average for 5 min). The clarified ribosomal supernatant is then layered in equal portions over a chilled discontinuous gradient of 10 ml of 0.5 M sucrose and 10 ml of 2.0 M sucrose, both buffered in TKMV. A swinging-bucket rotor is used in this instance (e.g., Spinco type SW-25.1). After layering the sample with a capillary pipet, the tubes are centrifuged overnight (at 1°C) at 25,000 rpm (54,000 X g average) for about 16 hr (5—7 X 10^7 g-min). Under these conditions the pellets contain mostly polyribosomes with very few monomers and subunits. The viscosity barrier at the 2.0 M sucrose interface retards the monomeric and dimeric ribosomes, while the residual dextran sulfate and detergent remain above in the 0.5 M sucrose buffer. The upper portion (up to the 2.0 M interface) is carefully removed with a suction pipet or syringe and discarded. The uppermost portion (about 1—2 ml of the interface) of the 2.0 M sucrose buffer can be removed, as some precipitated detergent and ferritin contaminate this region. The remaining

2.0 M sucrose buffer is decanted quickly and diluted with about 3—5 vol of TKMV buffer. These ribosomes are collected by centrifugation at about 6 X 10^7 g-min (100,000 X g average for 600 min). The 2X ribosomes from this region are predominantly monomers and dimers. The 2X polyribosome pellets can be suspended in 0.25 ml of TKMV per gram of original liver. They are opalescent when suspended, and the pellets are generally transparent. The yields of high polymeric ribosomes are in excess of 3 mg/g of original liver tissue and have an A_{260}/A_{280} ratio of about 1.6—1.7. Figure 2A illustrates a typical polyribosomal profile obtained from mouse liver. Polyribosomes and ribosomes may be stored in pellet form at -60°C to -100°C or, preferably, in a TKMV-G-G buffer for cell-free studies (at $\geq$200 A_{260} units/ml); this buffer contains 10 mM TES or HEPES (pH 7.5), 10 mM KCl, 1.2 mM $MgCl_2$, 0.1 mM EDTA, 20% (v/v) glycerol, and 10 mM glutathione (reduced).

VI. AMINO ACID INCORPORATION SYSTEM

The complete system for amino acid incorporation is a scaled-down modification of the system described for reticulocytes by Allen and Schweet [12]. The final volume of the reaction mixture is 0.35 ml, and either a mixture of [^{3}H]L-amino acid-reconstituted protein hydrolyzate or a mixture of 19 unlabeled L-amino acids minus the isotopic amino acid under study is used. Portions of 100 µl each are put on Whatman 3 MM filter paper discs and washed according to the procedure of Mans and Novelli [21].

One milliliter of the complete reaction mixture contains: 1—2 mg of ribosomes; 0.5—1.0 mg of AS_{70} (see Section VII); 50—100 μg of soluble RNA; 20—40 μg of creatine kinase; 0.7 μmole ATP; 35 μmoles tris—chloride (pH 7.5); 7 μmoles creatine phosphate; 13 μmoles GSH; 35 μmoles KCl; 3.5 μmoles $MgCl_2$; 0.17 μmole GTP; and 0.035 μmole of each amino acid. The chilled components are pipetted into tubes kept in an ice bath, and the $MgCl_2$, enzymes, ribosomes, and radioactive amino acids are the last four components, added in this order just before incubation. For small-volume incubations (i.e., 0.35 ml) a "mix" is prepared from concentrated solutions of the other components in order to conserve on volume and still have the required precision and accuracy. Also, we have routinely used similar amounts of phosphoenolpyruvate and pyruvate kinase in place of the creatine kinase energy-generating system.

VII. PREPARATION OF ENZYMES FOR PROTEIN SYNTHESIS

It is advantageous to have an enzyme preparation active for amino acid incorporation that can be stored for long periods without significant loss of activity and can be reused after repeated freezing and thawing. Such a preparation from reticulocytes or hepatic cells, a modification of that developed previously [12], is described below; it is known as AS_{70}. Largely because of the nuclease problem, an enzyme preparation from BHK cells is not recommended. The detergent-free supernatant, largely freed of ribosomes by centrifugation, is adjusted to pH 6.5 and 0.1 M tris with 2 M tris—HCl (pH 7.5) and 1 M acetic acid. Neutralized protamine sulfate solution (10 mg/ml) is added to a final concentration of 0.17 mg/ml. After 30 min the mixture is clarified by centrifugation and the protamine—RNA precipitate saved for preparation of RNA [22]. Powdered $(NH_4)_2SO_4$ is added to 40% saturation. After 30 minutes, the suspension is clarified by centrifugation at 10,000 $\times$ g for 15 min, the precipitate is discarded, and powdered $(NH_4)_2SO_4$ added to the supernatant to 70% saturation. After the suspension is allowed to stand for 30 min with occasional stirring, the precipitate is collected by centrifugation at 10,000 $\times$ g for 15 min. The precipitate is dissolved in 0.02 M tris buffer (pH 7.5) containing 1 mmole/liter GSH and 0.1 mmole/liter EDTA. The dialyzed enzyme fraction is stored at -20°C in the presence of 25 mmoles/liter GSH and 1 mmole/liter EDTA. This preparation is more stable and more active than the so-called "pH 5 enzyme" [12]. The AS_{70} from guinea pig liver is stable for over 6 months after the final concentration of the dialyzed enzyme is adjusted to 20% glycerol, 20 mmoles/liter GSH, 1 mmole/liter DTT, and 1 mmole/liter EDTA.

VIII. SUCROSE DENSITY GRADIENT CENTRIFUGATION

Linear sucrose gradients are prepared in a device similar to that described by Britten and Roberts [23]. The methodology for sucrose gradient analysis and fractionation of polyribosomes has been excellently detailed by Noll [24]. A gradient device manufactured by Buchler Instruments, Fort Lee, New Jersey, provides a means of preparing three gradients simultaneously when used with their four-channel peristaltic pump. Our data indicate that sucrose gradients vary in resolving power depending on the type of gradient and speed of centrifugation. Briefly, polyribosomes can be fractionated on a 10—30% linear sucrose gradient at 25,000 rpm in a SW-25.1 rotor for 2—4 hr. Better resolution is obtained with a 10—50% linear gradient and centrifugation for 16 hr at 12,000—15,000 rpm. Thus improved resolution is obtained when fractionating polyribosomes or other macromolecules at slower speeds in sucrose gradients with increased viscosity. All runs are made at 0—4°C. The gradients are fractionated with a device manufactured by Buchler Instruments that permits tube puncture at the bottom. The device is modified so as to permit gradient collection from the top. The screw-on top is increased in height and the inside fitted with a silicone rubber cone. An 18-gage needle is inserted and cemented in the vortex of the cone. Tygon tubing, with 1/32 in. inner diameter and 1/32 in. wall, is attached to the needle, which in turn leads to the flow cell. The screw-on cap is designed to permit the top of the centrifuge tube to fit snugly against the silicone rubber cone when the cap is properly tightened. A 60% sucrose solution is pumped into the bottom and the gradient is pumped out through a flow cell (Helma Cells, Inc., New York, New York). The flow cell is designed to permit bubbles to pass out of the cell. Absorbance can then be recorded using a Gilford recording spectrophotometer or similar instrumentation.

ABBREVIATIONS

Tris, tris(hydroxymethyl)aminomethane; GSH, reduced glutathione; EDTA, ethylenediaminetetraacetic acid; HEPES, N-2-hydroxyethylpiperazine-N'2-ethanesulfonic acid; TES, N-tris (hydroxymethyl)methyl-2-aminoethanesulfonic acid; DTT, dithiothreitol (Cleland's Reagent); SDS, sodium dodecyl sulfate.

REFERENCES

[1] M. Nomura, Bacteriol. Rev., 34, 228 (1970).

[2] C. Kurland, Science, 169, 1171 (1970).

[3] J. E. Darnell, Jr., Bacteriol. Rev., 32, 262 (1968).

[4] R. Traut, H. Delius, C. Ahmad-Zadeh, A. Bickle, P. Pearson, and A. Tissieres, Cold Spring Harbor Symp. Quant. Biol., 34, 25 (1969).

[5] B. Attardi, B. Cravioto, and G. Attardi, J. Mol. Biol., 44, 47 (1969)

[6] R. Lockard and J. Lingrel, Biochem. Biophys. Res. Commun., 37, 204 (1969).

[7] R. Ascione and R. Arlinghaus, Biochim. Biophys. Acta, 204, 478 (1970).

[8] S. Heywood, Nature, 225, 696 (1970).

[9] R. Miller and R. S. Schweet, Arch. Biochem. Biophys., 125, 632 (1968).

[10] J. Breillatt and S. R. Dickman, J. Mol. Biol., 19, 227 (1966).

[11] G. Blobel and V. R. Potter, Proc. Nat. Acad. Sci. USA, 55, (1966).

[12] E. Allen and R. S. Schweet, J. Biol. Chem., 237, 760 (1962).

[13] P. Marks, C. Wilson, J. Kruh, F. Gross, Biochem. Biophys. Res. Commun., 8, 9 (1962).

[14] R. Arlinghaus, R. Heintz, and R. S. Schweet, in Methods in Enzymology, Vol. 12, Part B (L. Grossman and K. Moldave, eds.), Academic Press, New York, 1968, p. 700.

[15] I. Macpherson and M. Stoker, Virology, 16, 147 (1962).

[16] B. S. Baliga, A. W. Pronczuk, H. N. Munro, J. Mol. Biol., 34, 199 (1968).

[17] H. Sidransky, M. Bongiorno, D. S. R. Sarma, and E. Verney, Biochem. Biophys. Res. Commun., 27, 242 (1967).

[18] S. H. Wilson, A. Z. Hill, and M. B. Hoagland, Biochem. J., 103, 567 (1967).

[19] R. L. Huston, L. E. Schrader, G. R. Honold, G. R. Beecher, W. K. Cooper, and H. E. Sauberlich, Biochim. Biophys. Acta, 209, 220 (1970).

[20] A. E. V. Haschemeyer and J. Gross, Biochim. Biophys. Acta, 145, 76 (1967).

[21] R. J. Mans and G. D. Novelli, Arch. Biochem. Biophys., 94, 48 (1961).

[22] E. Allen, E. Glassman, E. Cordes, and R. S. Schweet, J. Biol. Chem., 235, 1868 (1960).

[23] R. J. Britten and R. B. Roberts, Science, 131, 32 (1960).

[24] H. Noll, in Techniques in Protein Biosynthesis (P. N. Campbell and J. R. Sargent, eds.), Vol. 2, Academic Press, New York, 1969, p. 101.

Chapter 3

TISSUE CULTURE POLYRIBOSOMAL SYSTEMS

Richard Ascione, Ralph B. Arlinghaus,* and George F. Vande Woude

Plum Island Animal Disease Laboratory
Veterinary Sciences Research Division
Agricultural Research Service
U.S. Department of Agriculture
Greenport, New York

I. INTRODUCTION 60

II. EFFECT OF RNase INHIBITORS ON POLYRIBOSOME ISOLATION 66

III. EFFECT OF INCREASING CONCENTRATIONS OF DEXTRAN SULFATE ON ISOLATION OF POLYRIBOSOME PREPARATIONS. 67

IV. EFFECT OF CATIONS AND DETERGENTS ON THE YIELDS OF POLYRIBOSOMES. 70

V. INCUBATION CONDITIONS OF TISSUE CULTURE CELLS AND POLYRIBOSOMES. 74

VI. ISOLATION OF POLYRIBOSOMES FROM MAMMALIAN CELLS. . . 78

VII. VIRUS-INFECTED TISSUE CULTURE CELLS. 83

VIII. PREPARATION OF MEMBRANE-BOUND AND FREE RIBOSOMES . . 92

IX. CELL-FREE STUDIES WITH TISSUE CULTURE POLYRIBOSOMES. 96

X. REQUIREMENTS FOR CELL-FREE PROTEIN SYNTHESIS 100

XI. PREPARATION OF TISSUE CULTURE RIBOSOMAL SUBUNITS . .105

XII. PREPARATION OF RIBOSOME-ASSOCIATED I FACTORS FROM BHK POLYRIBOSOMES. 106

*Present address: M.D. Anderson Hospital and Tumor Institute, Houston, Texas.

XIII. INCORPORATION ASSAY WITH SUBUNITS AND I FACTOR . . .108
XIV. I-FACTOR-DEPENDENT BINDING OF N-ACETYL-L-PHE-tRNA. . 109
REFERENCES 112

I. INTRODUCTION

Many of the most important advances in molecular biology have arisen from studies employing bacterial cells as model systems. Apart from the relative simplicity of the bacterial cell, compared to the animal cell, the predominant motivation in investigating the bacterial cell was the ease of cultivation of large quantitites of homogeneous, single-cell organisms. This feature permitted the isolation and scrutiny of sufficient amounts of subcellular components. Furthermore, such bacterial cultures could be easily defined, especially with respect to their nutritional environment, and many biochemical studies were facilitated by the case with which the microenvironment of the laboratory cultivation system could be regulated.

Only recently, comparable studies have been made with animal cells by use of the tools of molecular biology. Understandably, because of its complexity, animal cell molecular biology has lagged behind studies with prokaryotes in culture. However, the past decade has witnessed a tremendous surge in animal cell molecular biology, attributable in part to tissue culture propagation methods that use well-characterized media under defined conditions.

The molecular biology of tissue culture systems, while still in its infancy, seems to offer many promising opportunities in the study of virology, oncology, immunology, pharmacology, and cellular differentiation. Numerous publications dealing with the application and varied uses of animal cells in culture cannot be enumerated in this chapter; we restrict ourselves to some of the tissue culture systems and methodology that have been developed for molecular biological investigations relevant to protein synthesis.

An animal cell is about 1000 times larger than the average *Escherichia coli* cell. For example, the dry weight of a single HeLa cell (the first human cell to be cultured continuously) is about 5×10^{-7} mg, compared to a single *E. coli* cell of about 2×10^{-10} mg. Nevertheless, in spite of this weight difference, the majority of enzymic reactions occur by almost identical pathways, and such a fundamental process as protein synthesis is functionally similar in all cells be they bacterial, plant, or animal.

Many tissue culture cell types have been utilized for studies of protein synthesis. Several cell lines have been established that retain their ability to synthesize specialized products. Thus fibroblastic cells in many cases still continue to make collagen. Also relatively simple to maintain continuously in culture are the tumor-derived specialized cells, many of which still continue to synthesize hormones or specific proteins.

Recently, many studies have concentrated on the subcellular sites of protein synthesis in tissue culture cells, the polyribosomes (or polysomes). These structures are organized clusters containing ribosomes, tRNA, and enzymes which simultaneously translate the information for sequencing peptides from the strand of mRNA, thus synthesizing in stepwise fashion numerous specific proteins. The first step in the formation of these polysomal structures appears to be the attachment of a nascent mRNA chain to the 40 and 60 S ribosomal subunits. This initiation process requires specific factors (see Chapter 6 of Volume I of this series). The initiation of protein synthesis has only recently been elucidated in animal cells [1-3]. Whether the initiation complex forms at or near the site of the nuclear membrane, where the mRNA departs from the nucleus, is as yet unknown. In any event the rapid association of many such ribosomes leads to the formation of polyribosomes.

Polyribosomes have been isolated from various cultured cells, among them baby hamster kidney (BHK) [4], Chang liver [5], Chinese hamster kidney [6], Ehrlich ascites tumor [7], HeLa (human cervical) [8], mouse L cells [9], and mouse fibroblast (3T6) cells [10]. Undegraded polyribosomes were quite easily obtained from some of these cells, while in many other cultured cells it has been almost impossible to recover undegraded polyribosomes. The isolation and quality of polyribosomes are apparently related to the presence of endogenous nucleolytic activity, as well as to the amount of membranous endoplasmic reticulum associated with such proteins. The ribosomes associated with the polyribosomal structures are fairly resistant to nuclease attack, especially in the presence of certain divalent cations. However, a single, unprotected mRNA strand that interconnects the ribosomes is extremely delicate and is highly sensitive to nucleolytic degradation, as well as to mechanical shearing forces. In BHK cells, for example, attempts to isolate polyribosomes by procedures used for rabbit reticulocytes [11] and HeLa cells [12] were unsuccessful largely because of the relatively higher amounts of endogenous nucleases and the high percentage of polyribosomes bound to membranes.

Although much of the rationale for the preparation of polyribosomes has already been discussed (see Chapter 2), in this chapter we detail more precisely the reasons for the methodology applied to various tissue culture systems.

A tissue culture sample containing 10^7 to 10^9 cells (approximately 10-1000 mg wet weight), preferably approximately 10^8 cells, is used for polyribosome analysis. If less than 10^7 cells are used, determinations of radioactivity in nascent polypeptides must be relied upon as the sole criterion for polysome analysis; absorbance recordings would be too insensitive. At the other extreme, a sample of 10^9 cells allows for analytical as well as some preparative polyribosomal studies.

Not all the polyribosomes obtained from a sample this large should be used in a single gradient for analysis, since too large a sample overloads the resolving capacity of most sucrose density gradients (via zone precipitation).

Cells in suspension cultures can be easily sampled; they are gently collected by centrifugation at 600–1000 × g for 3–5 min. Cells growing in mono- or multilayers can be removed from the vessel surface by brief treatment with trypsin; however, we do not recommend this method because most commercial grades of trypsin available for tissue culture work have RNase as a contaminant. Needless to say, EDTA (Versene) should not be used to remove cells from the vessel, since polyribosomes are destabilized in the presence of excess chelator. Rather, we prefer to use a modified rubber policeman to gently scrape the cells from the surface into an ice-cold liquid medium. For 2-liter, cylindrical, Baxter-type tissue culture bottles, we have constructed a spring-steel scraper having two standard windshield wipers with rubber blades. This scraper is used with several circular motions and this procedure is sufficient to gently remove all cells from the vessel surface.

Cells should be washed free of medium, especially those grown in the presence of nuclease-containing serum. Unfortunately, many tissue culture cells do not stand up to repeated washings; therefore a buffer containing 0.5% bovine serum albumin (fraction V) is best suited to adsorb and stabilize the tissue culture cells during the wash procedure. After gentle centrifugation the cells are suspended in a hypotonic buffer solution. Such buffers, provided they contain sufficient concentrations of stabilizing divalent cations, preserve polyribosomes and, at the same time, cause cells to swell, facilitating gentle lysis. It is known that strong, shearing mechanical forces disrupt polyribosomal structures; consequently, gentle homogenization is used. This can best be accomplished by a loose-fitting ball-type homogenizer as originally described by Dounce et al. [13]. Cells can be disrupted by as few as 5 strokes; however, 10 to 20 strokes do not seem to disrupt polyribosomes significantly, nor to lyse nuclear membranes (providing approximately 1–3 mM Ca^{2+} is present to stabilize nuclear membranes). The optimal time for suspension of cells in hypotonic buffer should be determined experimentally. For BHK cells 3 min is sufficient, while for ascites sarcoma cells at least 10 min is best. It should be emphasized that all preparative procedures are to be conducted as quickly as possible, and materials kept as close to 0°C as possible.

Unlike reticulocytes and mammalian organ cells, most tissue culture systems, particularly those amenable to virus infection studies, seem to demand special considerations. Generally, most tissue culture systems we examined seem to contain fairly high levels of endogenous RNase, as well as a rather high level of membrane-associated polyribosomes. Consequently, most routine isolation techniques that depend upon detergents to remove polysomes from membranes also cause activation of particle-bound (lysosomal) RNase. Although homogenization buffers of pH values of about 8 minimize to some extent the particulate RNase activity (optimum ca. pH 5), hydrolysis of the polyribosomal template strand occurs. Eliminating all detergents during preparation obviates such effects; however in the case of BHK cells, only about 15–20% of the total ribosomes can be recovered in the absence of detergents. Other cell types, such as HeLa, primary kidney cells, and reticulocytes, have only a small fraction of their total polyribosomes bound to endoplasmic reticulum, however. Thus, with these cells, a detergent-free isolation procedure might be preferred. Since our initial goal was to develop an isolation procedure for quantitative recovery of ribosomes from most tissue culture cells, the use of detergents was considered essential. Detergents were tested in conjunction with various nuclease inhibitors to determine which combinations yielded the best quantity and quality (with respect to polyribosome/monosome ratios).

II. EFFECT OF RNase INHIBITORS ON POLYRIBOSOME ISOLATION

Various nuclease inhibitors, such as bentonite and Zn^{2+}, were found to improve the quality of polyribosomes but to reduce the yields. Some of the other classic inhibitors of RNase tested were iodoacetate, carboxymethyl cellulose, and Cu^{2+}. When such inhibitors were included in the homogenization media (Figure 1B–D), the ratio of polyribosomes to monosomes was significantly increased compared to the homogenization procedure using reticulocyte standard buffer (RSB) with no RNase inhibitors (Figure 1A). Not surprisingly, the above-mentioned inhibitors gave polyribosomes that were inactive in subsequent cell-free studies. Most probably, the ribosomal proteins were inactivated by these inhibitors in the same manner as were the endogenous nucleases.

III. EFFECT OF INCREASING CONCENTRATIONS OF DEXTRAN SULFATE ON ISOLATION OF POLYRIBOSOME PREPARATIONS

Dextran sulfate has been shown to be a potent inhibitor of RNase [14]. When judiciously used in the homogenization medium, it was found to protect against polyribosome degradation (Figure 2). Furthermore, dextran sulfate-treated BHK polyribosomes were substantially more active in endogenous amino acid incorporation *in vitro* than untreated polyribosomes.

The concentration of dextran sulfate present during homogenization is very critical for the isolation of polyribosomes. Dextran sulfate concentrations in excess of 100 μg/ml result in partial disaggregation and deproteinization of polyribosomes. However, the ratio of polyribosomes to ribosomes increased in the range 1–100 μg/ml. Concentrations of dextran sulfate less than 20 μg/ml did not protect sufficiently against polyribosome disaggregation attributable to RNase, since an increase in small polyribosomes (monomers, trimers) was observed as the dextran sulfate concentration was lowered.

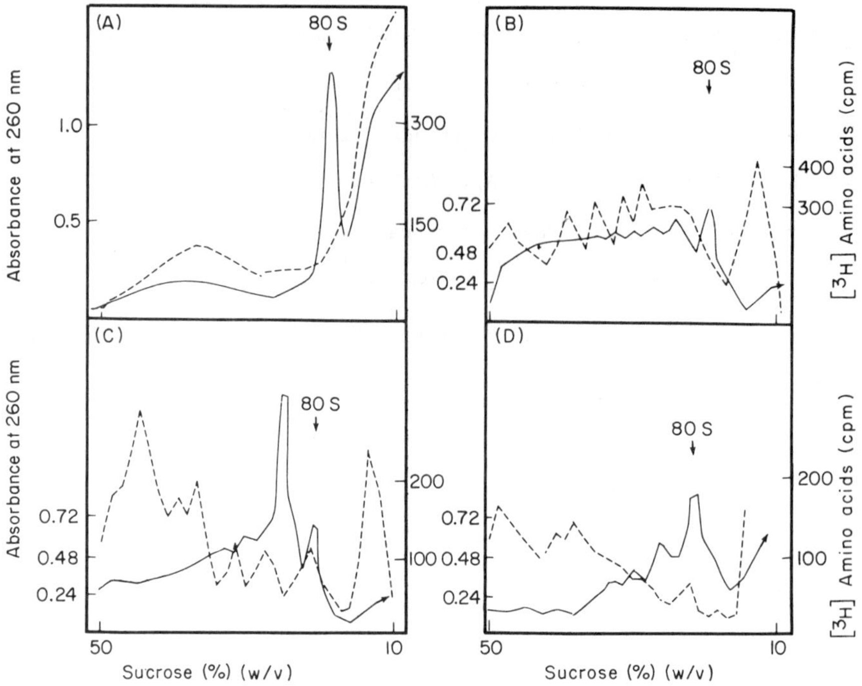

Fig. 1. Sucrose gradient centrifugation profile of BHK polyribosomes. Polyribosomes prepared using (A) reticulocyte standard buffer (RSB); (B) RSB containing 1 mM sodium iodoacetate; (C) RSB containing 300 μg/ml of carboxymethyl cellulose resin in suspension; (D) RSB containing 3 mM $CuCl_2$. BHK cells were pulsed <u>in vivo</u> for 3 min prior to harvest with 100 μCi of ^{3}H-reconstituted protein hydrolyzate (per Baxter bottle). Centrifugation and analysis of gradients as described in Chapter 2; 15,000 rpm, 16 hr in a Spinco SW-25.1 swinging-bucket rotor. Solid line, absorbance at 260 nm; dashed line, counts per minute of acid-insoluble [^{3}H]amino acids.

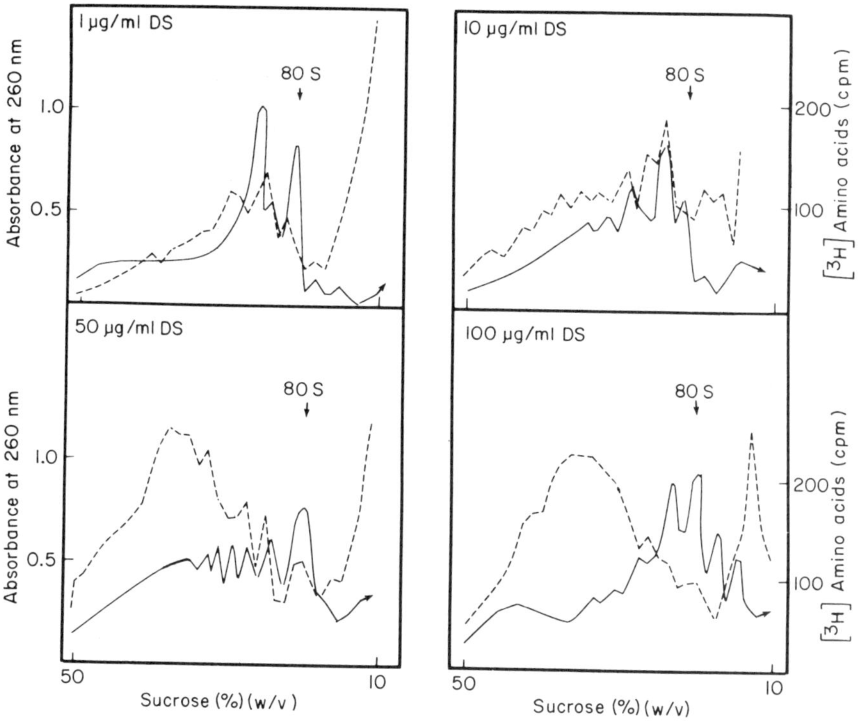

Fig. 2. Sucrose gradient centrifugation profiles of BHK polyribosomes prepared using various concentrations of dextran sulfate 500 (DS). Concentrations of dextran sulfate are indicated in upper left of each profile. Conditions of pulse labeling and centrifugation were as described in the legend to Figure 1. Solid line, absorbance at 260 nm; dashed line, counts per minute of acid-insoluble [^{3}H]amino acids.

Prolonged exposure of polyribosomes to dextran sulfate, even when the concentration was as low as 1 μg/ml, resulted in distortion of the polyribosomal profile. Therefore all preparations of BHK polyribosomes were washed by overnight centrifugation through a 2 M sucrose gradient buffered with 0.01 M tris (pH 7.5), 0.01 M KCl, 0.0012 M $MgCl_2$, and 0.0001 M EDTA (TKMV). These washed polyribosomes were stable even after storage at 4°C overnight; however, more prolonged storage resulted in gradual depolymerization of polyribosomes. More importantly, there was no appreciable loss of in vivo pulse-labeled [^{3}H]amino acids associated with the polyribosomes after resuspension in sucrose. There was a small but significant loss of endogenous, cell-free, amino acid-incorporating activity after such polyribosomes were frozen. In spite of the use of dextran sulfate to protect polyribosomes against nuclease degradation, all tissue culture polyribosomes isolated and washed as described above were depolymerized by additions of bovine RNase or EDTA (Figure 3).

IV. EFFECT OF CATIONS AND DETERGENTS ON THE YIELDS OF POLYRIBOSOMES

As expected, variation in cations present, as well as the presence of or absence of detergents, had a profound effect on the total yield of BHK ribosomes and polyribosomes (Table I). In the absence of nonionic detergent, particularly at low salt concentrations, most of the cytoplasmic ribosomes cosedimented with the nuclear and cytoplasmic membranes at low centrifugal forces (10,000 × g for 15 min or less). In fact, as mentioned previously, homogenization in the absence of nonionic detergent gave a postmitochondrial supernatant fluid that contained less

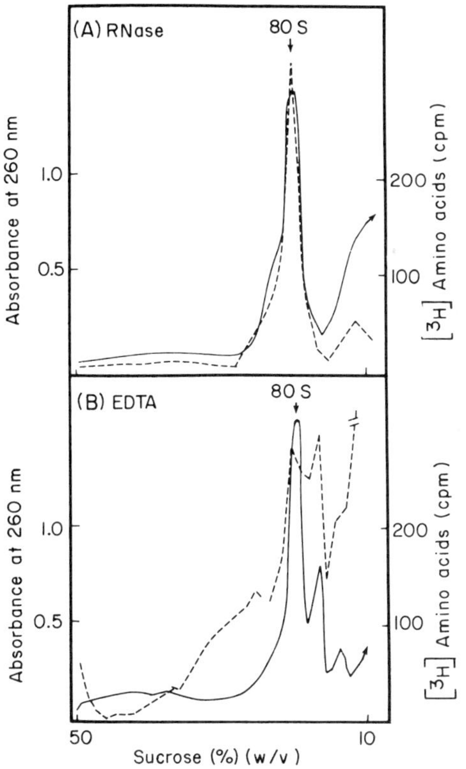

Fig. 3. Effect of RNase and EDTA on BHK polyribosomes prepared in the presence of 50 μg/ml of DS. (A) RNase: 2X polyribosomes (in TKMV) incubated for 10 min at room temperature with 0.1 μg/ml of crystalline bovine pancreatic RNase and then layered on sucrose gradient for analysis. (B) EDTA: 2X polyribosomes (in TKMV) adjusted to 2 mM EDTA and layered on sucrose gradient for analysis. All other conditions were as described in the legend to Figure 1. Solid line, absorbance at 260 nm; dashed line, counts per minute of acid-insoluble [^{3}H]amino acids.

TABLE I

Effect of Cations and Detergent on Yield of Polyribosomes

Homogenization conditions		Ribosome yield[a]	Analysis of sucrose gradients		
Monovalent/divalent ratio (mM)	Nonionic detergent (%)	(mg/10^9 BHK cells)	<Dimers (%)	Dimers to pentamers (%)	>Pentamers (%)
10:1	—	1.55	65	22	13
10:1	0.5	4.05	72	18	9
150:15	—	2.35	24	34	41
150:15	0.5	10.00	25	30	45
150:15	1.0	12.0	27	26	47
500:50	—	6.10	33	18	49
500:50	0.5	11.05	36	9	55

[a] 1 mg/ml ribosomes = 11.8 A_{260} units.

than 15% of the total ribosomes of a BHK cell. Therefore, to increase the yield of ribosomes in the postnuclear fraction, one of several types of nonionic detergents was used at concentrations of 0.5 and 1.0%. The resultant increase in ribosomes was a result of the solubilization of membrane-associated ribosomes. Once the nuclei were removed, ribosomal membranes still present as microsomes were further solubilized by the addition of deoxycholate (DOC) to a 0.5% concentration. Even with dextran sulfate present, only a small fraction of the solubilized ribosomes were found as polyribosomes at low cation concentrations. Thus the polyribosomes were stabilized against degradation by increasing the $MgCl_2$ and KCl concentrations 5- to 10-fold higher than those used for reticulocyte polyribosomes (in conjunction with dextran sulfate). To prevent aggregation or dissociation of ribosomes, the ratio of monovalent to divalent cations was always maintained at approximately 10:1 [15].

Once the polyribosomes were isolated from the cytoplasm and rinsed free of detergent by recentrifugation, they were found to be quite stable, even in gradients that contained as little as 0.1 mM $MgCl_2$. In fact, divalent cation concentrations greater than 3 mM Mg^{2+} appeared to cause considerable aggregation of BHK polyribosomes unless the concentration of K^+ was increased concomitantly.

V. INCUBATION CONDITIONS OF TISSUE CULTURE CELLS AND POLYRIBOSOMES

To isolate polyribosomes from tissue culture cells, the metabolic state of the cells must be considered and controlled. One factor that markedly influences the distribution of ribosomes in cells is the phase of cell growth. Early stages (1—3 days) in the growth of tissue-cell cultures (Figure 4A)

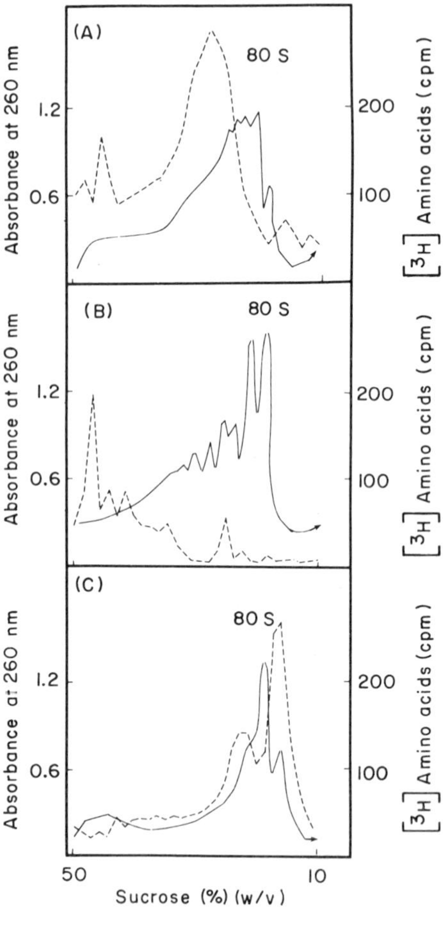

yield a higher proportion of ribosomes as polyribosomes than do later stages (10—13 days)(Figure 4B).

In addition to growth phase, the tissue culture medium also greatly influences the proportion of polyribosomes to ribosomes. Deprivation of amino acids, or even one essential amino acid, for 30 min or more results in a pronounced disaggregation of polyribosomes (Figure 4C). Therefore, if polyribosomes are to be labeled with amino acids, cells should be briefly pulsed immediately after they are rinsed free of growth medium; they should never be subjected to prolonged preincubation in the absence of amino acids. Polyribosomes similarly dissaggregate to monoribosomes when induced by glucose starvation. This starvation disaggregation has been previously reported to affect polysomal distribution in various mammalian organ cells [16], as well as in ascites tissue cells [7].

Fig. 4. Polyribosomes isolated from BHK cells after varied growth conditions. BHK cells grown: (A) 1-3 days in rolling Baxter bottles; (B) 10-13 days in rolling Baxter bottles; and (C) in medium containing only nonessential amino acids for 30 min prior to harvest. All pulse-labeling and centrifugation conditions were as described in the legend to Figure 1; 50 μg/ml of dextran sulfate were used during polyribosome preparation. Solid line, absorbance at 260 nm; dashed line, counts per minute of acid-insoluble [^{3}H]amino acids.

The rate at which tissue culture cells are cooled during harvesting must also be considered. If cells are cooled rapidly by the addition of a mixture of frozen and liquid buffer solution, the proportion of ribosomes in polyribosomes is quite high (Figure 5A). However, if cells are harvested in

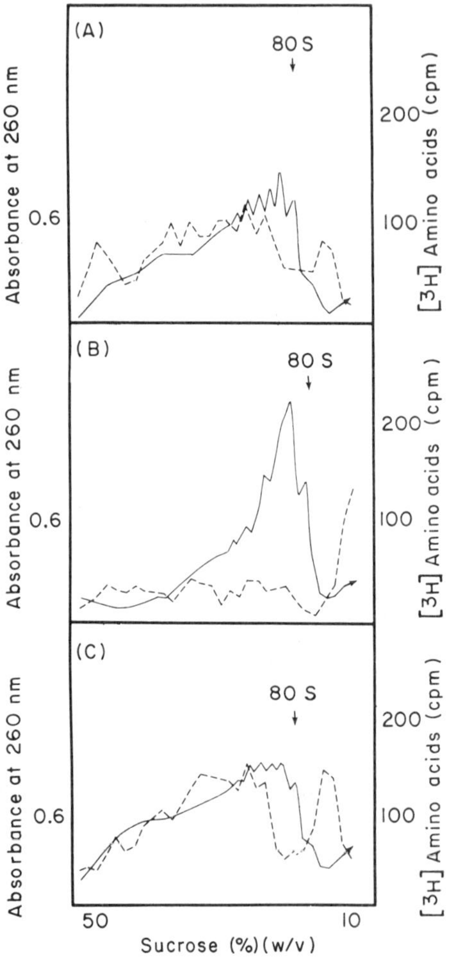

a tray of ice with only cold buffer (stored at 4°C) as the suspension medium, the proportion of ribosomes in polyribosomes is substantially decreased (Figure 5B). Note the concomitant increase in the absorbance of monosomes, while no increase is seen in the amount of labeled polypeptides associated with these monosomes. An increase in radioactivity of the monosomes would be expected if nuclease activity were responsible for degradation of the polyribosomes during cooling. Interestingly, addition of cycloheximide (Figure 5C) to the harvesting buffer (kept at 4°C) prevented the breakdown of polyribosomes. Presumably cycloheximide inhibits the translational movement of ribosomes ("runoff") along the mRNA [17] (see Vol. 1, Chapter 1) that may take place during the harvesting of the cells at 4°C or above. This runoff phenomenon is one of the more important

---

Fig. 5. Effect of cell harvest conditions on BHK polyribosomes. (A) Cells were rapidly chilled (<30 sec) and scraped using buffer containing iced buffer (at 0°C). (B) Cells were slowly chilled (>1 min) using cold buffer (4°C). (C) Cells were showly chilled (>1 min) using cold buffer (4°C) containing 2 mM cycloheximide. All pulse-labeling and centrifugation conditions were as described in legend to Figure 1; 50 μg/ml of dextran sulfate were used during polyribosome preparation. Solid line, absorbance at 260 nm; dashed line, counts per minute of acid-insoluble [^{3}H]amino acids.

parameters that must be controlled to maintain the distribution of polyribosomes in tissue culture cells that are metabolically active.

VI. ISOLATION OF POLYRIBOSOMES FROM MAMMALIAN CELLS

Several tissue culture cells not usually amenable to the isolation of polyribosomes have had their polyribosomes prepared by the dextran sulfate—detergent technique. Polyribosomes were isolated from primary calf kidney, porcine kidney, and mouse sarcoma (ascites) as shown in Figure 6A—D. In addition,

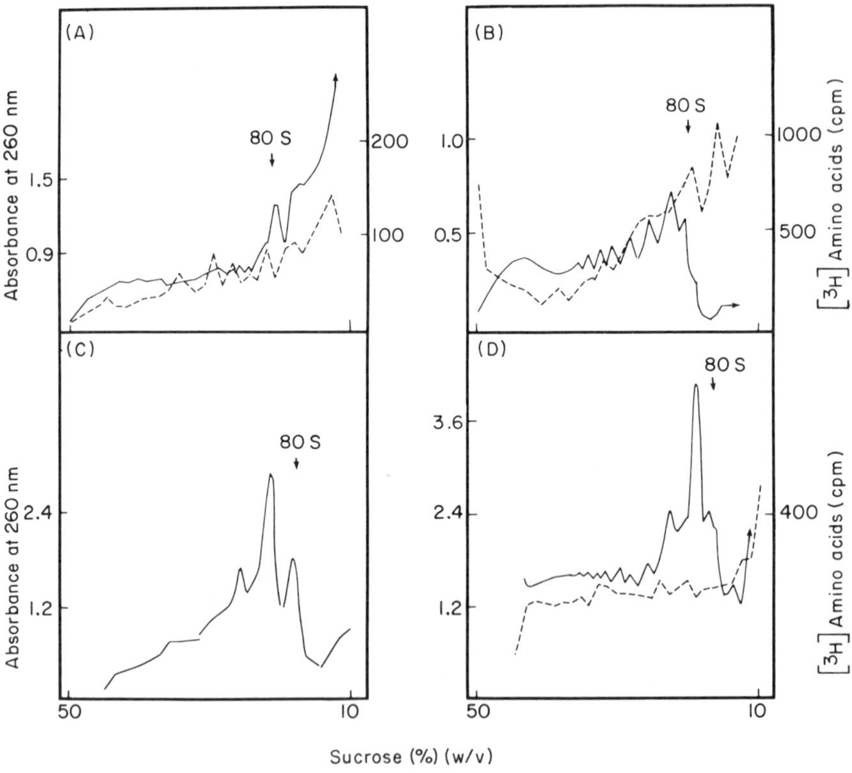

polyribosomes were isolated from nonconfluent primary monolayer cultures of calf kidney cells and pig kidney cells (Figure 7A and B). Note that the specific activity of the ^{3}H-labeled polypeptide associated with polyribosomes is very much reduced in confluent monolayer cultures (Figure 6A and B), as compared with nonconfluent cultures (Figure 7A and B). This might reflect the rapid switch-off of metabolic functions as a result of contact inhibition. No such effect is noted in multilayered tissue culture cells (those not subject to contact inhibition, such as BHK and HeLa cells).

The following is a generalized procedure for the extraction of polyribosomes from tissue culture cells with the aid of

---

Fig. 6. Preparation of polyribosomes from various tissue culture cells. (A) Primary calf kidney cells (confluent monolayer cultures). (B) Primary porcine kidney cells (confluent monolayer cultures). (C) Mouse ascites sarcoma 180 cells (frozen and thawed in ascitic fluid plus 10% dimethyl sulfoxide). (D) Bovine turbinate cells (confluent monolayer cultures, seventh passage). All pulse-labeling and centrifugation conditions were as described in the legend to Figure 1; 50 μg/ml of dextran sulfate were used during polyribosome preparation. Solid line, absorbance at 260 nm; dashed line, counts per minute of acid-insoluble [^{3}H]amino acids.

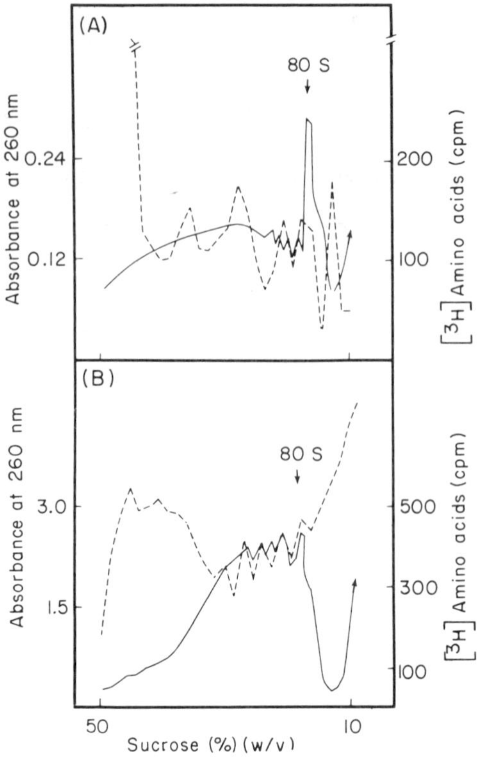

Fig. 7. Preparation of polyribosomes from nonconfluent cell monolayer cultures in: (A) primary porcine tissue culture cells; (B) primary bovine tissue culture cells. All pulse-labeling and centrifugation conditions were as described in the legend to Figure 1; 50 μg/ml of dextran sulfate were used during polyribosome preparation. Solid line, absorbance at 260 nm; dashed line, counts per minute of acid-insoluble [^{3}H]amino acids.

an RNase inhibitor. All cells, if desired, can be appropriately pulse labeled prior to harvest with amino acids (for 3 min or less) to label the nascent proteins associated with the polyribosomes. Approximately 10^8 to 10^9 tissue culture cells are gently scraped from bottles and washed free of growth medium with the aid of 100 ml of ice-cold,buffered, bovine serum albumin solution containing cycloheximide (20 mM tris—HCl, 150 mM KCl, 5 mM $MgCl_2$, 1 mM cycloheximide, 5 mg/ml of bovine serum albumin fraction V),the pH is adjusted to 7.5 at 20°C; cells are recovered by centrifugation at low speed. The cell pellet is resuspended with a Dounce-type homogenizer at 0°C in 5 ml of hypotonic buffer solution [20 mM tris—HCl, (pH 7.5), 60 mM KCl, 3.6 mM $CaCl_2$, 5.0 mM $MgCl_2$] and allowed to swell with intermittent agitation with a vortex mixer for 3—10 min depending upon the cell type and its resistance to breakage (e.g., kidney culture cells need approximately 3—5 min swelling time, whereas ascites-type cells may require 10 min. Immediately after an equal volume of hypertonic buffer containing nonionic detergent [40 mM tris—HCl (pH 7.5) 2.0% Triton X-100 (Rohm and Haas) or NP-40 (Shell Chemical Company), 0.60 M sucrose, 0.25 M KCl, 10 mM $MgCl_2$, 3.6 mM $CaCl_2$, 1 mM EDTA] is added, the cell slurry is mixed and homogenized with 5 to 10 strokes with a loose-fitting pestle. Depending upon intracellular nuclease levels (i.e., the cell type), immediately before homogenization a RNase inhibitor such as dextran sulfate 500 (Pharmacia Fine

Chemicals) may be added to the slurry at a final concentration of 50 μg/ml. Other RNase inhibitors can also be used with somewhat less effectiveness such as the 100,000 × g supernatant fluid from rat liver (Ref. [18] and Chapter 4) or yeast RNA [19]. These natural inhibitors are suitable for many cell-free preparations from tissue culture cells that are to be utilized for subsequent enzymic studies with which dextran sulfate could interfere. The lysed cells are centrifuged to remove cell debris and nuclei at 1000 × g for 5 min, and the cytoplasmic fraction is carefully removed. The yield of polyribosomes from cells with high amounts of membraneous endoplasmic reticulum can be increased by resuspending the nuclei and debris once again in 5 ml of a 1:1 mixture of hypotonic buffer and hypertonic buffer solutions containing nonionic detergent and homogenizing five more times. After centrifugation at 1000 × g for 5 min, the supernatant fraction is combined with the previous cytoplasmic fraction and adjusted to 10 mM $MgCl_2$ and 5 mM $CaCl_2$ to stabilize the polyribosomes against degradation prior to the addition of DOC to a final concentration of 0.5%. The detergent-treated cytoplasm is centrifuged at 100,000 × g or faster for 60—90 min to produce a translucent polyribosomal pellet. These polyribosomes, designated 1×, are rinsed by decantation with 5 ml of a 10 mM tris—HCl (pH 7.5)——10 mM KCL——1 mM $MgCl_2$——0.1 mM EDTA (TKMV) buffer and are resuspended with the aid of a hand-operated Teflon pestle in 1—2 ml of TKMV buffer. The suspension is layered onto a 30-ml discontinuous gradient composed of 10 ml of 0.5 M TKMV-buffered sucrose, 15 ml of 1.1 M TKMV-buffered sucrose, and 5 ml of 2.0 M TKMV-buffered sucrose. It is then centrifuged overnight (at least 20 hr) at 63,000 × g in a swinging-bucket-type (e.g., Spinco SW-25.1) rotor. For smaller preparations a smaller-scale, 5-ml gradient can be utilized (in a Spinco SW-39 rotor) and centrifuged at 84,000 × g for 16—17 hr. The pelleted polyribosomes are designated 2× (washed) polyribosomes and can be analyzed on a sucrose gradient or stored frozen as a pellet at -60°C for cell-free incorporation studies.

VII. VIRUS-INFECTED TISSUE CULTURE CELLS

Studies of tissue culture polyribosomes are particularly useful when applied to virus-infected cells. Depending upon the cell used, a particularly abrupt and well-characterized alteration in cellular metabolism occurs subsequent to picornavirus infection. Host cell protein synthesis declines rapidly [20,21], and many of the cellular polyribosomes are dissociated into monomers and dimers shortly after infection [22]. After the appearance of virus-specific RNA, a species of polyribosomes appears that is active in the synthesis of nascent viral proteins [23,24]. In general, the virus-infected tissue culture cells are excellent for the study of viral RNA biosynthesis [25,26], virus particle maturation, viral interference with host protein synthesis (including the maturation of host ribosomes and its interruption [27]), and control of the translation of the viral message [28].

We have isolated decigram quantities of purified foot-and-mouth disease virus (FMDV) [29]; a rolling Baxter bottle was used to grow BHK tissue culture cells on a mass scale [30]. Such a system has facilitated investigations of protein synthesis in virus-infected cells and has clearly shown distinctive differences in the polyribosomes of infected cells as compared with the distribution of polyribosomes in uninfected host cells.

An example of alteration of host protein synthesis after FMDV infection is depicted in Figure 8A—C. Infected tissue culture cells were pulsed with [^{3}H]amino acids to label the nascent polypeptides associated with the functional polyribosomes. Polyribosomes were then isolated as described previously using dextran sulfate and then partitioned on a sucrose gradient. At 1½ hr after infection, the absorbance patterns of the polyribosomes do not appear to be visually distinct from those of the control cell polyribosomes. However, the presence of ^{3}H-labeled nascent polypeptides on polyribosomes is appreciably inhibited in the infected cells relative to the controls. Nascent polypeptide labeling in infected cells becomes extensive only after 3 hr and is primarily associated with polyribosome species with sedimentation rates of >400 S. Since the predomi-

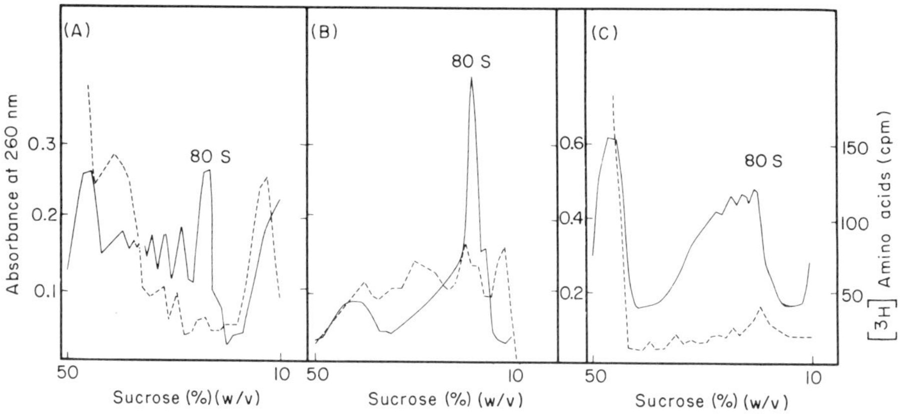

Fig. 8. Effect of FMDV infection on BHK polyribosomes. (A) Noninfected (control) cells. (B) BHK cells infected for 90 min with type-A-119 FMDV at a multiplicity of infection of 1-10 pfu per cell. (C) BHK cells infected for 260 min with type-A-119 FMDV at 5-10 pfu per cell. All gradients contained a 5-ml cushion of 50% sucrose containing 20% Ficoll. Pulse-labeling and centrifugation conditions were as described in the legend to Figure 1; 50 μg/ml of dextran sulfate were used during polyribosome preparation. Solid line, absorbance at 260 nm; dashed line, counts per minute of acid-insoluble amino acids.

---

nant labeled nascent polypeptides in the uninfected cell are associated with polyribosomes sedimenting at <400 S, it seems that the size of the messages is basically different from that of the infected cell polyribosomes. Not surprisingly, this size distribution suggests that the nascent polypeptides synthesized subsequent to FMDV infection are different from those of the host cell. Differences in nascent polypeptides of the host and infected cells can be determined by sodium dodecyl sulfate (SDS)—acrylamide gel electrophoresis [31]. The SDS—polyacrylamide electrophoretic method of Summers and Maizel [32] for the estimation of the number and molecular weights of polypeptides can be usefully applied to the analysis of nascent polypeptides associated with polyribosomes. Thus it has been possible to study the synthesis, to estimate the number and molecular weights of _in vivo_ labeled products programmed by viral RNA [33], and to resolve the distribution of polypeptides associated with the various cytoplasmic polyribosomes (i.e., membraneous and free) as a function of infection.

As an example of nascent polypeptide analysis on SDS—polyacrylamide gels, cells were infected for 3 hr with FMDV, in the presence of 5 μg/ml of actinomycin D (present for 2 hr prior to harvest to inhibit host mRNA production), and then pulsed with (1—2 mCi/10^8 cells) isotopic amino acids for 2—3 min. The label was chased for specified times (into completed polypeptides) by the addition of an excess of unlabeled amino acids. During isolation of the polyribosomes, it is essential that small polypeptides (ranging from oligomeric peptides of molecular weight $\leq$5000) be removed since they are not properly sieved on SDS gels, thus causing a spurious background of counts. The usual polyribosome isolation procedure is sufficient to remove the bulk of the (labeled) small polypeptides and free amino acids. However, if a high background of counts is present, a 10% trichloroacetic acid (TCA) wash, followed by an 80% acid-acetone wash, is recommended [34]. An aliquot of polyribosomes, containing at least 10,000 cpm and about 200 μg or less of protein, should be boiled for 1—2 min in 2% SDS and 2% β-mercaptoethanol to remove nascent polypeptides,which are electrophoresed as recommended by Maizel et al. [34]. Typical polyacrylamide gel profiles for polypeptides obtained from FMDV-infected cells are illustrated in Figure 9 A—D. The nascent polypeptides associated with the infected cell polyribosomes (and microsomes) (Figure 9A) are generally larger and more heterogeneous than the completed polypeptides released into the postmicrosomal supernatant liquid (Figure 9C). More than 10 virus-specific polypeptides can be resolved on polyacrylamide gels using in vivo labeled proteins extracted from FMDV-infected cell polyribosomes. The highest of the nascent-polypeptide molecular weights appeared greater than 100,000 daltons (Figure 9A, gel slices 1 to 15). These large nascent polypeptides are thought to be viral polypeptide precursors since they are absent in polyribosomes that have been pulsed and than chased (Figure 9B). Note, however, that several newer and smaller-molecular-weight

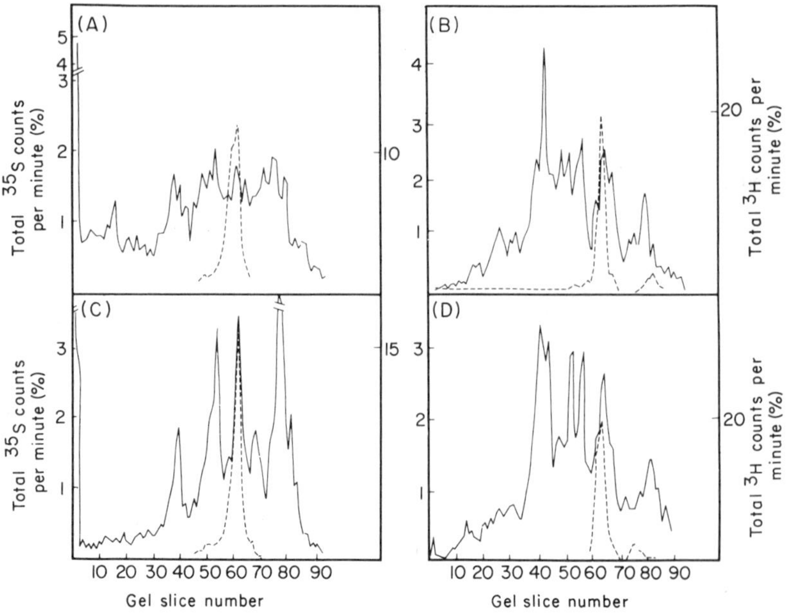

Fig. 9. Polyacrylamide electropherograms of in vivo labeled proteins obtained from FMDV-infected cells pulse labeled 150 min after infection. Solid line, [^{35}S]methionine-labeled protein obtained from: (A) polyribosomes, 3 min pulse labeled; (B) polyribosomes, 3 min pulse labeled and chased 45 min with excess unlabeled methionine; (C) postmicrosomal supernatant, 3 min pulse labeled; (D) postmicrosomal supernatant, 3 min pulse labeled and chased 45 min with excess unlabeled methionine. Cell-infected samples were coelectrophoresed with [^{3}H]amino acid-labeled A-119 virion protein (dashed line).

components appeared subsequently (slices 70 and 80). These latter species appear to remain associated with the polyribosome fraction and are not released into the postmicrosomal supernatant even after a 45-min chase (Figure 9D). The size and distribution of the large nascent polypeptides as a function of isotope kinetics suggests that a posttranslational cleavage mechanism may be present as with other picornavirus polypeptides [34].

The production of virus-specific polyribosomes in infected cells requires certain special conditions [35]. It is best to obtain a nearly synchronous infection by using a multiplicity of infection of at least 1 to 10 (i.e., 1 to 10 infectious particles per cell). Actinomycin D (5 μg/ml) is also helpful, as it suppresses the synthesis of host cell mRNA and accentuates the synthesis of virus products. Also, the polyribosomes of BHK cells should be harvested no later than 3—4 hr after infection with FMDV since nuclei subsequently become fragile and lyse easily. When this occurs, it is necessary to reduce the number of homogenizations by one-half.

Many virus-infected cells appear to synthesize virus-specific proteins preferentially on polyribosomes associated with the membraneous endoplasmic reticulum. Thus subfractionation of the polyribosomes into membrane and free species (see Section VII for procedure) should reveal significant differences with respect to the distribution of virus-specific proteins.

Indeed, significant differences do exist between the size distributions of membrane-bound and free polyribosomes in an uninfected BHK cell (Figure 10A and B). In the uninfected cell both ribosomal species show an almost equal specific activity with respect to nascent polypeptide labeling. By contrast, the specific activity of nascent proteins associated with free polyribosomes and isolated 210 min after FMDV-infection is drastically reduced. Not surprisingly, however, the specific activity of nascent proteins associated with

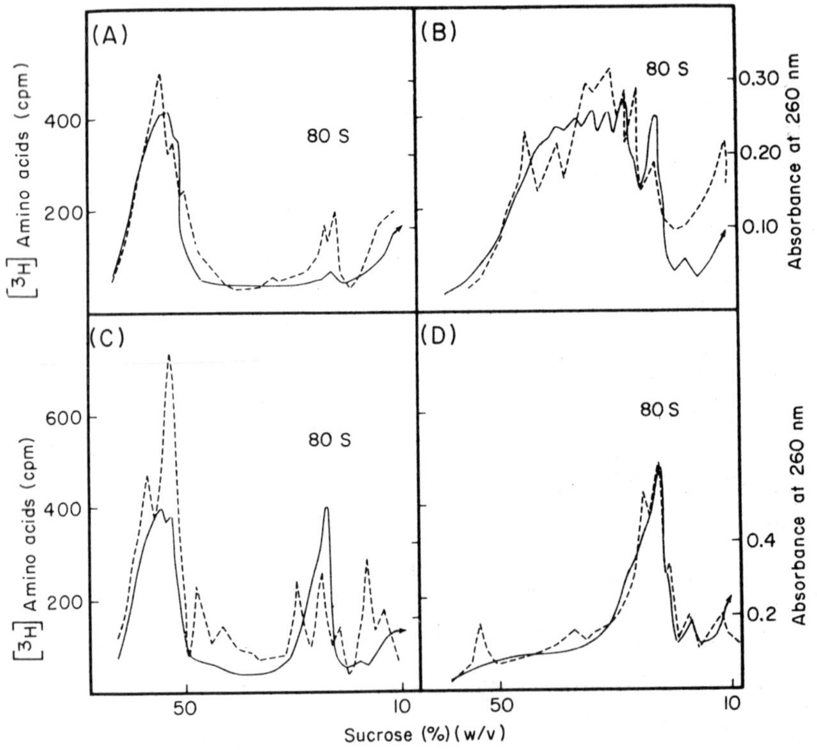

membrane-bound polyribosomes is enhanced by 210 min after infection, consistent with a stimulation of virus-specific protein synthesis. While the overall protein synthesis in the FMDV-infected cell appears to be inhibited, particularly as reflected by the total polyribosome distributions (see Figure 8), certain species of polyribosomes are actually stimulated relative to the control (e.g., the membrane-bound species). Therefore many different parameters of polyribosome distribution and protein synthesis should be examined prior to making interpretations concerning virus-induced alterations in infected tissue culture cells.

Fig. 10. Effect of FMDV infection on free and membrane-bound polyribosomes isolated from BHK cells. (A) Membrane-bound ribosomes prepared from noninfected cells. (B) Free polyribosomes prepared from noninfected cells. (C) Membrane-bound polyribosomes from FMDV-infected cells. (D) Free polyribosomes from FMDV-infected cells. All cells were infected with 10 pfu of FMDV A-119 virus and incubated for 240 min. Gradients contained a 5-ml cushion of 60% sucrose containing 10% Ficoll. Pulse-labeling and centrifugation conditions were described in the legend to Figure 1; 50 μg/ml of dextran sulfate were used during polyribosome preparation. Solid line, absorbance at 260 nm; dashed line, counts per minute of acid-insoluble [^{3}H]amino acids.

VIII. PREPARATION OF MEMBRANE-BOUND AND FREE RIBOSOMES

Several procedures can be used to fractionate bound and free ribosomes without first isolating the total ribosome population. For liver tissue these procedures separate subcellular components, in the absence of any detergent, by differential centrifugation and then separate the different ribosomal species by zonal centrifugation [36]; advantage is taken of the fact that the density of membrane-bound ribosomes is lower than that of free ribosomes [37]. The quantities of free and bound ribosomes can be determined by difference from the total population of ribosomes obtained after treatment with detergent [38]. Usually, as much as 35% of the membrane-bound ribosomes remain associated with the nuclei and are generally discarded with the premitochondrial fraction (the material sedimented after centrifugation for 150,000 g-min). Consequently, we describe a procedure for isolating large polyribosomal complexes associated with the premitochondrial membranes as well as those sedimenting with nuclear membranes.

A typical procedure for BHK cells is as follows (other tissue culture cells can be treated similarly). Approximately 3-5 X 10^8 BHK cells (about 0.5 gm wet weight) are rapidly cooled and removed from tissue culture bottles. The cells are washed free of growth medium by resuspension in about 100 vol of isotonic tris buffer solution containing 0.16 M tris (pH 7.5, at 25°C), 5 mM $MgCl_2$, 3.6 mM $CaCl_2$, and 1 mM EDTA and centrifugation at low speed. The cell pellet is resuspended in a hypotonic sucrose buffer containing 10 mM tris (pH 7.5), 50 mM KCl, 3 mM $MgCl_2$, 2 mM $CaCl_2$, 0.5 mM EDTA, and 0.25 M sucrose using 2.5-3.0 vol of solution per gram of wet weight of cells. After 10 min in ice (periodically stirring on a vortex mixer), the slurry is hand-homogenized 20 to 30 times in a loose-fitting Dounce homogenizer. This procedure is gentle enough to lyse cells without fragmenting the nuclei. Sufficient homogenization, however, is required to fragment endoplasmic reticulum only partially. Vigorous and extensive homogenization fragments the endoplasmic reticulum into small microsomal components without a substantial increase in the amount of free ribosomes [38]. After 10-20 μg/ml of dextran sulfate 500 are added to the lysed homogenate, the intact nuclei and cell debris are removed by low-speed centrifugation at 2000 X g for 5 min. The nuclear pellet can be saved for further use. Meanwhile, the supernatant fluid is sedimented at 100,000 X g for 60 min to obtain a detergent-free microsomal pellet which is then gently resuspended in 1-2 ml of TKMV buffer and layered over 30 ml of a linear, 10-65% (w/v) sucrose gradient buffered in TKMV (at 1°C). After centrifugation overnight at 40,000 X g (average) (16,000 rpm) for 16 hr in a swinging-bucket rotor (e.g., Spinco SW-25.1), the membrane-bound polyribosomes are separated from the free polyribosomes. Figure 11 illustrates that the membrane-bound polyribosomes are located in the 54%

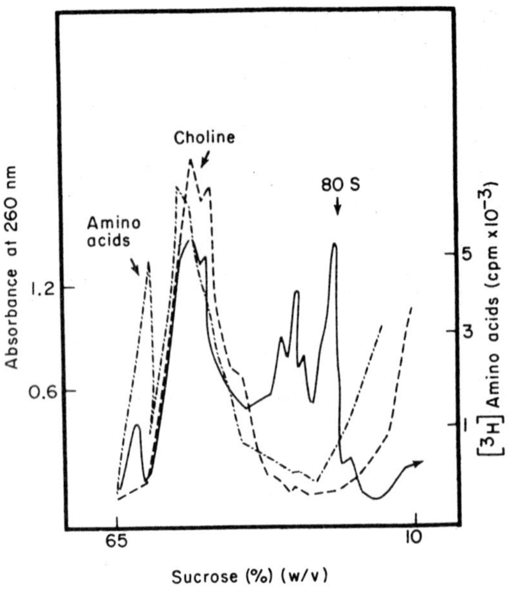

Fig. 11. Preparation and analysis of membrane-bound and free BHK polyribosomes isolated in the absence of detergent. Solid line, absorbance at 260 nm; dashed line, counts per minute of [^{14}C]choline label after incubating cells in presence of 20 μCi per bottle of [^{14}C]choline for 3 hr; dotted and dashed lines, counts per minute of acid-insoluble [^{3}H]amino acids incubated in the presence of 200 μCi per bottle of ^{3}H-reconstituted protein hydrolyzate for 5 min. Gradients of 10-65% TKMV buffered sucrose were centrifuged at 13,500 rpm for 16 hr in a Spinco SW-25.1 swinging-bucket rotor.

(w/v) sucrose region of the gradient (seen associated with labeled choline). Note that the free monomeric and oligomeric ribosomes sediment in the upper half of the gradient, while the free polyribosomes are almost pelleted. It appears that the specific activity (with respect to amino acid pulse labeling) of free polyribosomes in BHK cells is comparable to that of membrane-bound species. Moreover, membrane-bound polysomes comprise approximately 60-70% of the polyribosomes in the gradient.

The separated nuclear pellets have been found to contain approximately 25-35% of the total polyribosomes in association with nuclear membranes. These polysomes can be recovered by resuspending the pellets at least twice more in the detergent-free hypotonic sucrose solution; after homogenization and centrifugation of this fraction, the supernatant fluid can be combined with the original (postnuclear) microsome supernatant. However, the pelleted nuclei may be suspended in 10 vol of nonionic detergent-sucrose (nuclei wash) buffer and homogenized five times to remove any cytoplasmic components containing membrane-associated ribosomes. The nuclei wash buffer consists of 10 mM tris (pH 7.5), 150 mM KCl, 5 mM $MgCl_2$, 3 mM $CaCl_2$, 0.8 mM EDTA, 0.25 M sucrose, 1% Triton X-100 (or Lubrol WX), and 25 μg/ml dextran sulfate 500. After homogenization the washed nuclei are pelleted by low-speed centrifugation (5000 X g for 5 min) and the supernatant is diluted with 5 vol of nuclei wash buffer adjusted with DOC to a 0.5% final concentration and then centrifuged for 100,000 X g for 60 min. The pelleted nuclear membrane-associated polyribosomes arc resuspended in TKMV (1-2 ml) and layered over 25 ml of a linear 10-50% (w/v) sucrose (and a 5-ml cushion containing 60% sucrose) plus 10% Ficoll gradient. All sucrose gradients are buffered with TKMV and are centrifuged overnight (in a Spinco SW-25.1 rotor). The nuclear membrane-associated polyribosomes are resolved by centrifuging for 25,000 X g (13,500 rpm) for 16 hr (Figure 12).

IX. CELL-FREE STUDIES WITH TISSUE CULTURE POLYRIBOSOMES

While most cell-free protein synthesis studies have centered upon reticulocyte and liver cell preparations, largely because these cells yielded relatively undegraded and functional polyribosomes [39], a few tissue culture systems have yielded polyribosomes amenable to cell-free protein synthesis studies. However, the extent of in vitro endogenous (native) amino acid incorporation in the latter systems appeared to be related to the method used to isolate the polyribosomes and the extent of degradation. Accordingly, polyribosomes isolated in the absence of dextran sulfate appeared to have a significantly lower level of endogenous activity than polyribosomes prepared with dextran sulfate (Table II). The removal of dextran sulfate from in vitro amino acid incorporation experiments is essential, however, since the addition of dextran sulfate at concentrations used for polyribosome isolation is definitely inhibitory (Table II). Furthermore, several workers have

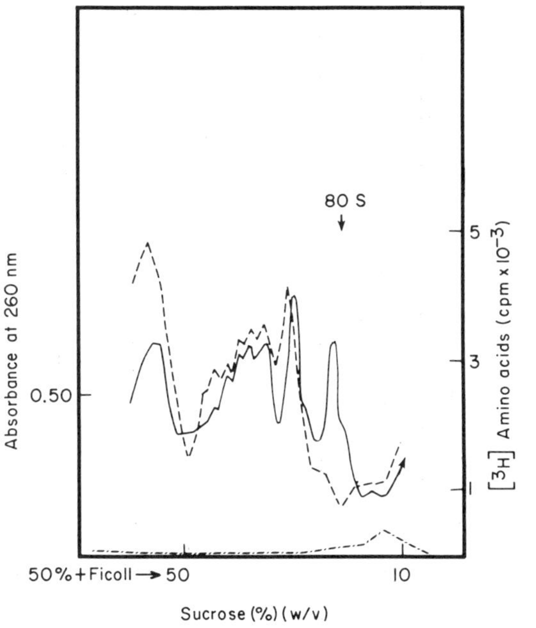

Fig. 12. Nuclear membrane-associated membrane-bound polyribosomes isolated after detergent treatment of BHK nuclei. Solid line, absorbance at 260 nm; dashed line, counts per minute of acid-insoluble [^{3}H]amino acid; dotted and dashed line, [^{14}C]choline label. Samples were layered on a gradient with a 5-ml cushion of 50% sucrose and 20% Ficoll and centrifuged as in Figure 11. All assay reaction mixtures contained 10 mM magnesium acetate, guinea pig liver supernatant factors, and 10 μCi/ml of [^{14}C]L-phenylalanine (specific activity, 25); 100-μl aliquots were removed at times indicated.

TABLE II

Endogenous Amino Acid Incorporation by Cell-Free Systems[a]

Ribosome source	Ribosome isolation procedure	Supernatant enzyme source	Assay additions	$[^{3}H]$L-Amino acid mixture	$[^{14}C]$L-Amino acid mixture	$[^{3}H]$L-valine	$[^{14}C]$L-phenylalanine
BHK	-	BHK	0	0.048	0.030	---	---
BHK	+	BHK	0	0.210	0.600	---	---
BHK	-	Mouse liver	0	0.296	---	0.047	0.023
BHK	+	Mouse liver	0	1.276	---	0.283	0.139
		Reticulocyte	0	1.300	---	0.275	0.148
Reticulocyte	-	Mouse liver	0	---	---	1.320	0.660

Reticulocyte	-	Mouse liver	Plus 50 μg/ml dextran sulfate	---	---	0.420	0.110
BHK	+	Mouse liver	Plus 50 μg/ml dextran sulfate	---	---	0.133	0.055
BHK	+	Mouse liver	Plus 50 μg/ml dextran sulfate	---	---	0.028	0.023

[a]+, With 50 μg/ml of dextran sulfate present during homogenization; -, without dextran sulfate.

Incorporation is expressed as nanomoles of amino acid(s) incorporated per milligram of ribosomes per hour at 37°C.

already shown that dextran sulfate [40] and other similar polyanions [41] inhibit protein synthesis by binding to the 30 S subunit of E. coli ribosomes. Recently, Korner [42] suggested that dextran sulfate is able to inhibit amino acid incorporation specifically by interfering with the initiation step. Obviously, to avoid either the introduction of dextran sulfate or nucleases into the cell-free assay, a nuclease-containing supernatant enzyme should not be used. Therefore supernatant enzymes derived from any of several sources should be used for optimal activity. Mouse liver and rabbit reticulocyte supernatant enzymes (see Chapter 1 for preparation) were similarly active in a heterologous system with BHK polyribosomes, and both enzymes were substantially more active than the homologous BHK enzyme (Table II).

X. REQUIREMENTS FOR CELL-FREE PROTEIN SYNTHESIS

BHK polyribosomes are absolutely essential for activity, as are Mg^{2+} and an energy source. There is also a high dependence upon supernatant enzymes and the presence of 19 L-amino acids, as well as on glutathione. The presence of pancreatic RNase and high concentrations of dithiothreitol (DTT) is inhibitory (Table III). The time course for the incorporation of amino acids into protein by BHK polyribosomes is shown in Figure 13A. Incorporation is linear for the first 15 min at 37°C, but continues to rise gradually thereafter for the next 45 min. Figure 13B shows that BHK polyribosomes

TABLE III

Requirements of Polyribosomal System from BHK Cells

Assay system	[^{14}C]Leucine incorporated[a]	Percent of control
Complete	0.360	100
Minus 10^5 X g supernatant	0.038	11
Minus BHK polyribosomes	0.005	1
Minus ATP-generating system	0.027	8
Minus ^{12}C (19 amino acids)	0.144	40
Minus glutathione (16 mM)	0.066	18
Plus DTT (20 mM)	0.023	6
Plus RNase (10 μg/ml)	0.030	8

[a]Incorporation is expressed as nanomoles per milligram of ribosomes per hour.

are extremely sensitive to the Mg^{2+} concentration. Maximal activity occurred at 6-7 mM Mg^{2+}; raising and lowering the concentration of this cation caused a marked inhibition of amino acid incorporation. With respect to the above requirements, the BHK cell-free system is similar to those described by other workers [43]. Surprisingly, the cell-free polyribosomal system from BHK tissue culture cells was not stimulated by additions of polyuridylic acid (poly U), an artificial mRNA, which directs the synthesis of polyphenylalanine (see Chapter 1; Volume 1). However, after the polyribosomal system was allowed to exhaust itself (i.e., after 60 min of incubation in

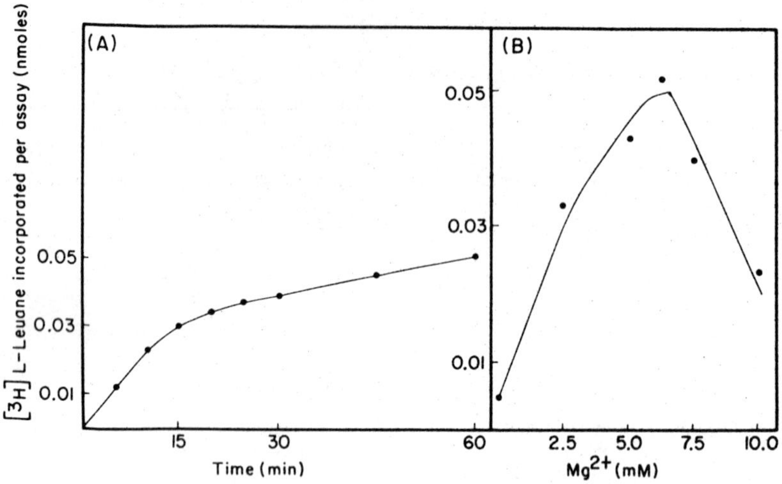

Fig. 13. Cell-free amino acid-incorporating system using 2X BHK polyribosomes with guinea pig liver supernatant factors. (A) Kinetics of [^{3}H]L-leucine incorporation using a scaled-down (100-μl) version of the cell-free assay described in Chapter 2 containing 10 μCi/ml of [^{3}H]L-leucine (specific activity, 100). (B) Incorporation of [^{3}H]L-leucine per 100 μl of assay as a function of magnesium acetate concentration when incubated at 37°C for 60 min.

---

a complete cell-free system), the ribosomes (mostly monomeric) responded to additions of synthetic (exogenous) messengers even though the endogenous incorporation was nearly nil. Moreover, adding a natural message such as FMDV RNA to the preincubation ribosomal system (while increasing the Mg^{2+} concentration to ≥10 mM) resulted in a marked stimulation in the incorporation of amino acids (Table IV). In a further

TABLE IV

Effect of Adding Exogenous RNA to BHK Polyribosomes[a]

System	[^{14}C]L-Phe incorporated	[^{3}H]Amino acid mix incorporated
Incorporation system		
Minus poly U (native)	0.165	0.638
Plus poly U (160 μg/ml)	0.078	0.159
Plus FMDV RNA (50 μg/ml)	0.095	0.498
Preincubated system		
Minus poly U (native)	0.006	0.048
Plus poly U (160 μg/ml)	0.950	---
Plus FMDV RNA (50 μg/ml)	---	1.390

[a]Incorporation is expressed as nanomoles of amino acid(s) per milligram of ribosomes per hour at 37°C.

investigation of this exogenously stimulated system, BHK polyribosomes were dissociated by treatment with a high concentration of salt (see Section XI and Chapter 1),by a modification of the procedure of Miller and Schweet [44]. Such salt "shocking," or dissociation, of polyribosomes yielded a stimulated system that was dependent upon a factor(s) removed from the polyribosomes by the salt treatment. Significantly, the incorporation system functioned at a low Mg^{2+} concentration ≤ 5 mM, and addition of exogenous mRNA (artificial or natural) stimulated the rate of amino acid incorporation 4- to 5-fold above the endogenous rate (Figure 14).

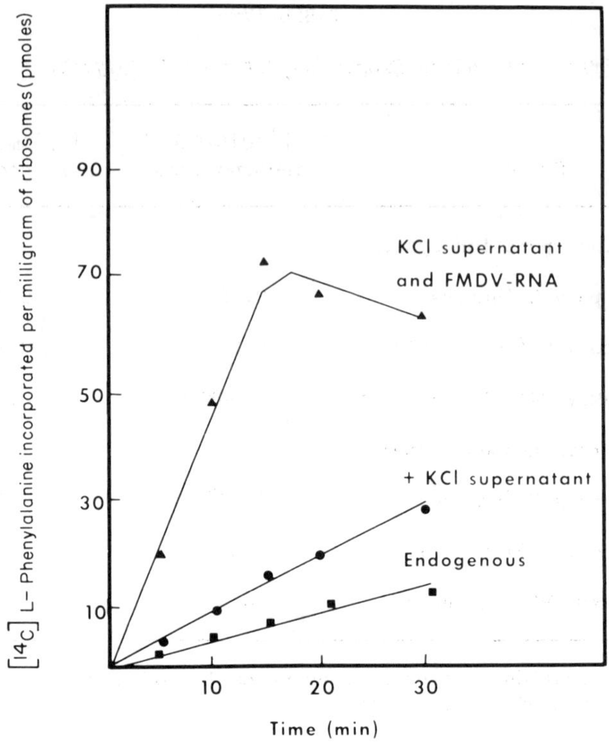

Fig. 14. Kinetics of incorporation of [^{14}C]L-phenylalanine using KCl-dissociated BHK ribosomes. Squares, endogenous refers to the activity observed using KCl-dissociated ribosomes only. Circles, KCl supernatant refers to that fraction remaining in the supernatant after centrifugation at 105,000 X g for 2-4 hr. Triangles, FMDV RNA prepared from purified virions and used at a concentration of 150 μg/ml.

It appears from these studies that intact, noninfected BHK polyribosomes are not well suited for incorporation studies with exogenous mRNA. Instead, investigations with a cell-free amino acid incorporation system dependent upon added template, such as viral RNA, should deal with dissociated polyribosomes and, in particular, active ribosomal subunits.

XI. PREPARATION OF TISSUE CULTURE RIBOSOMAL SUBUNITS

Ribosomal subunits have been implicated in the initiation of _de novo_ biosynthesis of viral and mammalian proteins [45, 46]. Several procedures can be employed to dissociate ribosomes into subunits, but obtaining biologically active subunits of mammalian ribosomes has been more difficult than obtaining active subunits from bacterial ribosomes. Merely lowering the Mg^{2+} concentration of the buffer in which bacterial ribosomes are suspended to 1 mM induces the formation of biologically active subunits, whereas most mammalian ribosomes remain intact at Mg^{2+} concentrations as low as 0.1 mM. It is possible by carefully adjusting the solutions to a proper K^+/Mg^{2+} ratio (in the presence of a sulfhydryl group protective agent) to dissociate (reversibly) mammalian ribosomes into subunits capable of biological activity. Several methods have been described for mammalian ribosomes, such as exposure to moderate concentrations of urea [47], or incubation of ribosomes with puromycin [48], or just exposure to high concentrations of monovalent salts [49].

The following procedure is a generalized one for the

preparation of ribosomal subunits that retain biological activity. The tissue culture polyribosomes, since they are in culture bottles, can be readily converted to monoribosome species by first allowing the cells to cool slowly (over a 120-min period) down to 4°C (in complete growth medium). After slow cooling most of the (BHK) polyribosomes are converted to monomers and can be easily dissociated once they have been isolated as described in the absence of dextran sulfate. The 1 X ribosomes are suspended in dissociation buffer consisting of 0.01 M tris (pH 7.5), 0.610 M KCl, 0.001 M $MgCl_2$, 0.0001 M EDTA, 0.002 M DTT (at 200 A_{260} units/ml) and are then incubated at 37°C for 15 min. After cooling, the aggregates are removed by low-speed centrifugation (10,000 X g for 5 min) and the dissociated ribosomes are collected by ultracentrifugation at 100,000 X g for 600 min. These ribosomes can be stored in a buffer consisting of 30 mM TES or HEPES (pH 7.5), 60 mM KCl, 1 mM $MgCl_2$, 1 mM DTT, 0.1 mM EDTA, and 20% glycerol at -60 to -100°C.

XII. PREPARATION OF RIBOSOME-ASSOCIATED I FACTORS FROM BHK POLYRIBOSOMES

In order to use the ribosomal subunits in a cell-free system, it is necessary to isolate the protein factors dissociated from polyribosomes. Since these crude factors (called I factors [44]) should be free of subunit contamination and have low endogenous messenger activity, it is necessary to use slightly different procedures to isolate them from polyribosomes than to dissociate the ribosomes into subunits. If the concentration of Mg^{2+} is increased during the dissociation step, the endogenous messenger activity of the I factors is reduced and, further, these factors appear to be more stable than the factors isolated from reticulocyte ribosomes. After attempts at column purification, however (e.g., on DEAE-cellulose), the I-factor activity becomes quite unstable; this problem is quite similar to that involving the behavior of reticulocyte I factors after purification [50].

To isolate the crude I factors from BHK or other tissue culture cells, it is best to start with polyribosomes rather than preincubated ribosomes, and preferably polyribosomes isolated in the absence of dextran sulfate. This polyanion, even at low concentrations, inhibits the I-factor-dependent incorporation system.

To dissociate the I factors from polyribosomes, the latter are suspended at 200 A_{260} units/ml in a solution containing 30 mM TES or HEPES (pH 7.5), 100 mM KCl, 10 mM $MgCl_2$, and 1 mM EDTA; 3 M KCl is added dropwise to a concentration of 850 mM, DTT is added to 2 mM (or GSH to 20 mM), and glycerol is added to 10%. The suspension is then stirred overnight (12-16 hr) at 1-4°C. The supernatant fluid, containing I factor, is isolated by centrifugation at 100,000 X g for 300 min and the pellets are discarded. The factors can be concentrated by adding 2.5 vol of saturated (1°C) $(NH_4)_2SO_4$ solution; after 30 min in ice the precipitate is collected by centrifugation (20 min at 15,000 X g) and dissolved in a minimum volume of dialysis fluid. The dialysis fluid consists of 10 mM TES or HEPES (pH 7.5), 60 mM KCl, 3 mM $MgCl_2$, 1 mM DTT, 0.3 mM EDTA, 30% glycerol, and 0.01-0.03 mM GTP. Dialysis is carried out for several hours at 4°C against two 500-ml changes of buffer.

The dialyzed I factor remains relatively stable for periods as long as 6 months, especially if it is stored in small aliquots at -20°C or less.

XIII. INCORPORATION ASSAY WITH SUBUNITS AND I FACTOR

The I factor and ribosomal subunits can be used to incorporate labeled Phenylalanyl-tRNA (Phe-tRNA) into polyphenylalanine in the presence of poly U at Mg^{2+} concentrations less than the 10 mM required when I factor is not present. This lowered requirement for Mg^{2+} in the presence of I factors has been noted by several workers [1,3] and is reminiscent of the lowered Mg^{2+} requirement for the incorporation of N-acetylphenylalanyl-tRNA (N-acetyl-Phe-tRNA) when bacterial initiation factors are used [51].

The poly U-directed polyphenylalanine synthesis assay is performed as follows, in a total volume of 1.0 ml of 50 mM pH 7.5 tris buffer containing 30-50 mM KCl, 3-5 mM $MgCl_2$, 1.0 mM DTT, 1.0 mM ATP, 0.5 mM GTP, 3.5 mM creatine phosphate,

33 μg/ml of creatine phosphokinase, 50-100 μg/ml of low-molecular-weight poly U (about 2-5 X 10^5 daltons), 15-30 A_{260} units of ribosomal subunits, 400-500 μg/ml of I factor, 800-1500 μg/ml of protamine-treated supernatant enzyme (liver or reticulocyte), and 10-20 A_{260} units of ^{14}C- or ^{3}H-L-Phe-tRNA [55]. For incorporation in the presence of subunits, the low-molecular weight poly U species is more active at the concentrations specified; with the larger polymer (about 10^6 molecular weight), 350-500 μg/ml of poly U are required for comparable activity. The reaction is linear for the first 15 min and levels off after 30 min of incubation at 30°C. An aliquot of the reaction (100 μl) is applied to a Whatman 3 MM filter paper disc (2.5 cm) which is soaked in 10% TCA (20 ml per disc). The filter paper is then placed in fresh 10% TCA and heated at 90°C for 15 min to hydrolyze the tRNA. After a further wash with cold 5% TCA, the filter disc is soaked (10 ml per disc) first in alcohol-ether (2:1) and then in ether. The filter is air dried and the entrapped polyphenylalanine determined in a liquid scintillation spectrometer.

XIV. I-FACTOR-DEPENDENT BINDING OF N-ACETYL-L-PHE-tRNA

The I factors and subunits can also stimulate the binding of N-acetyl-Phe-tRNA in the presence of poly U [52] and are therefore unlike the mammalian peptide chain elongation factors described earlier [53] which can not utilize N-acetylated

Phe-tRNA in binding to ribosomes at low Mg^{2+} [54].

The binding assay, in 200 μl, is as follows: 30 mM tris (pH 7.5) buffer, 40 mM KCl, 4 mM $MgCl_2$, 1 mM DTT, 0.4 mM EDTA, 50 μg of poly U, 1.0-2.0 A_{260} units of subunits, 50-100 μg/ml $AS_{67\%}$ I factor, 0.2 mM GTP, and 2-3 A_{260} units of [^{3}H] or [^{14}C]N-acetyl-L-Phe-tRNA acetylated by the procedure of Haenni and Chappeville [56]. The reaction is linear only during the first 3 min of incubation at 30°C. After completion the reaction is diluted with 20 vol of ice-cold buffer solution [20 mM tris (pH 7.5), 40 mM KCl, 4 mM $MgCl_2$, 0.4 mM EDTA], filtered through a micromembrane filter (Millipore or the like), and washed twice more with 40 vol of the wash buffer. The filters, containing the labeled acetylated tRNA-ribosome complexes, are dried and the radioactivity is determined. The binding system, similarly to the incorporation system, also shows I-factor dependence only at low Mg^{2+} concentrations. Unacetylated Phe-tRNA can be used in this binding assay, but because of the presence of contaminating elongation factors in crude I-factor preparations, little I-factor-dependent stimulation is observed. The extent of binding is usually expressed as picomoles of N-acetyl-Phe-tRNA bound per A_{260} unit or per milligram of ribosomes. Puromycin, which is often used to dissociate polyribosomes, has been avoided since it gives rise to variable binding in cell-free studies. Glycerol and DTT have been used during dissociation since these reagents appear to afford a measurable stabilizing influence on the normally labile I factors from ribosomes.

Thus far, the procedures described here have been used successfully to isolate I factors and subunits from various mammalian tissues (liver, kidney, and tissue culture cells). It remains to be determined whether or not the cell-free synthesis of mammalian proteins can be accomplished with the isolated subunits, I factors, and a specific natural mRNA. There is some indication that such a system, when used with a natural viral mRNA (FMDV RNA), is able to direct the cell-free synthesis of animal virus proteins [57].

ABBREVIATIONS

Tris, tris(hydroxymethyl)aminomethane; GSH, reduced glutathione; EDTA, ethylenediaminetetraacetic acid; HEPES, N-2-hydroxyethylpiperazine-N'-2-ethanesulfonic acid; TES, N-tris(hydroxymethyl)-2-aminoethanesulfonic acid; DTT, dithiothreitol (Cleland's reagent); SDS, sodium dodecyl sulfate; ME, 2-mercaptoethanol.

ACKNOWLEDGMENTS

We wish to thank Dr. Howard L. Bachrach for his critical evaluation of this article and, in addition, to acknowledge his support of our research by providing the necessary samples of purified virus and BHK cells. We also thank Dr. John H. Graves for supplying primary tissue culture cells and bovine turbinate continuous cell cultures. We are indebted to Dr. John McVicar for providing us with mouse ascites sarcoma 180 cells. The combined able assistance of Mrs. Cathy Loper, Mrs. Elaine Winters, Mr. Robert Moore, and Mr. Paul Nichols is also gratefully acknowledged.

REFERENCES

[1] R. Miller, Ph. D. Thesis, Univ. of Kentucky, 1967.

[2] A. E. Smith and K. A. Marcker, Nature, 226, 607 (1970).

[3] D. A. Shafritz, P. M. Prichard, J. M. Gilbert, and W. F. Anderson, Biochem. Biophys. Res. Commun., 38, 721 (1970).

[4] R. Ascione and R. B. Arlinghaus, Biochim. Biophys. Acta, 204, 478 (1970).

[5] E. Eliasson, G. E. Bauer, and T. Hultin, J. Cell Biol., 32, 287 (1967).

[6] D. L. Steward, J. R. Shaeffer, and R. M. Humphrey, Science, 131, 791 (1968).

[7] W. J. W. van Venrooij, E. C. Henshae, and C. A. Hirsch, J. Biol. Chem., 245, 5947 (1970).

[8] T. Borun, M. Scharff, and E. Robbins, Biochim. Biophys. Acta, 149, 302 (1967).

[9] L. Prevec and A. F. Graham, Science, 154, 522 (1966).

[10] B. Goldberg and H. Green, J. Mol. Biol., 26, 1 (1967).

[11] J. R. Warner, P. Knopf, and A. Rich, Proc. Natl. Acad. Sci. U.S., 49, 122 (1963).

[12] S. Penman, Y. Becker, and J. E. Darnell, J. Mol. Biol., 8, 541 (1964).

[13] A. L. Dounce, R. F. Witter, K. J. Monty, S. Pate, and M. A. Cottone, J. Biophys. Biochem. Cytol., 1, 139 (1955).

[14] S. R. Dickman, Science, 127, 1392 (1958).

[15] J. Breillatt and S. R. Dickman, J. Mol. Biol., 19, 227 (1966).

[16] M. Sidransky, M. Bongiorno, D. S. R. Sarma, and E. Verney, Biochem. Biophys. Res. Commun., 27, 242 (1967).

[17] A. R. Williamson and R. Schweet, J. Mol. Biol., 11, 358 (1965).

[18] G. Blobel and V. R. Potter, Proc. Natl. Acad. Sci. U.S., 55, 1283 (1966).

[19] A. E. V. Haschemeyer and J. Gross, Biochim. Biophys. Acta, 145, 76 (1967).

[20] E. F. Zimmerman, M. Heeter, and J. E. Darnell, Virology, 19, 400 (1963).

[21] G. F. Vande Woude, J. Polatnick, and R. Ascione, J. Virol., 5, 458 (1970).

[22] S. Penman, K. Scherrer, Y. Becker, and J. E. Darnell, Proc. Natl. Acad. Sci. U.S., 49, 654 (1963).

[23] D. F. Summers, J. V. Maizel, and J. E. Darnell, Proc. Natl. Acad. Sci. U.S., 54, 505 (1965).

[24] J. E. Darnell, M. Girard, D. Baltimore, D. F. Summers, and J. V. Maizel, in The Molecular Biology of Viruses (J. S.

Colter and W. Paranchych, eds.), Academic Press, New York, 1967, 375.

[25] J. M. Bishop, D. F. Summers, and L. Levintow, Proc. Natl. Acad. Sci. U.S., 54, 1237 (1965).

[26] R. B. Arlinghaus, H. L. Bachrach, and J. Polatnick, Biochim. Biophys. Acta, 161, 170 (1968).

[27] R. Ascione and G. F. Vande Woude, J. Virol., 4, 727 (1969).

[28] M. Willems and S. Penman, Virology, 30, 355 (1966).

[29] H. L. Bachrach, Natl. Cancer Inst. Monograph, 29, 73 (1968).

[30] J. Polatnick and H. L. Bachrach, Appl. Microbiol., 12, 368 (1964).

[31] J. V. Maizel, Jr., in Fundamental Techniques in Virology (K. Habel and N. P. Salzman, eds.), Academic Press, New York, 1969, 334.

[32] D. F. Summers and J. V. Maizel, Proc. Natl. Acad. Sci. U.S., 59, 966 (1968).

[33] J. A. Mudd and D. F. Summers, Virology, 42, 328 (1970).

[34] J. V. Maizel, Jr., D. F. Summers, and M. D. Scharff, J. Cell Physiol., 76, 273 (1970).

[35] R. Arlinghaus, J. Polatnick, and G. F. Vande Woude, Virology, 30, 541 (1966).

[36] C. M. Redman, J. Biol. Chem., 244, 4308 (1969).

[37] M. Roumiantzeff, M. V. Maizel, and D. F. Summers, Virology, in press.

[38] G. Blobel and V. R. Potter, J. Mol. Biol., 26, 279 (1967).

[39] R. Arlinghaus, R. Heintz, and R. Schweet, in Methods in Enzymology (L. Grossman and K. Moldave, eds.), Vol. 12, Part B, Academic Press, New York, 1968, 700.

[40] F. Miyazawa, O. R. Olijnyk, C. J. Tilley, and T. Tamaoki, Biochim. Biophys. Acta, 145, 96 (1967).

[41] T. Sinozawa, I. Yahara, and K. Imahori, J. Mol. Biol., 36, 305 (1968).

[42] A. Korner, Biochim. Biophys. Acta, 174, 351 (1969).

[43] A. Perani, B. Parisi, L. DeCarli, and O. Ciferri, Biochim. Biophys. Acta, 161, 223 (1968).

[44] R. L. Miller and R. Schweet, Arch. Biochem. Biophys., 125, 632 (1968).

[45] R. Ascione and R. Arlinghaus, Federation Proc., 28, 433 (1969).

[46] J. O. Bishop, Biochim. Biophys. Acta, 119, 130 (1966).

[47] M. L. Peterman, A. Pavlovec, and I. B. Weinstein, Federation Proc., 28, 725 (1969).

[48] T. E. Martin, F. S. Rolleston, R. B. Low, and I. G. Wool, J. Mol. Biol., 43, 135 (1969).

[49] A. K. Falvey and T. Staehelin, J. Mol. Biol., 53, 1 (1970).

[50] P. M. Prichard, J. M. Gilbert, D. A. Shafritz, and W. F. Anderson, Nature, 226, 511 (1970).

[51] M. Salas, M. Hille, J. Last, A. Wahba, and S. Ochoa, Proc. Natl. Acad. Sci. U.S., 57, 387 (1967).

[52] J. Lucas-Lenard and F. Lipmann, Proc. Natl. Acad. Sci. U.S., 57, 1050 (1967).

[53] R. B. Arlinghaus, J. Shaeffer, J. Bishop, and R. Schweet, Arch. Biochem. Biophys., 125, 604 (1968).

[54] J. Siler and K. Moldave, Biochim. Biophys. Acta, 195, 130 (1969).

[55] R. B. Arlinghaus, J. Shaeffer, and R. Schweet, Proc. Natl. Acad. Sci. U.S., 51, 1291 (1964).

[56] A. L. Haenni and F. Chappeville, Biochim. Biophys. Acta, 114, 135 (1966).

[57] R. Ascione and G. F. Vande Woude, Biochem. Biophys. Res. Commun., 45, 14 (1971).

Chapter 4

POLYRIBOSOMES AND CELL-FREE PROTEIN SYNTHESIS IN THE SPLEEN

Norman Talal

Department of Health, Education, and Welfare
Public Health Service
National Institutes of Health
Bethesda, Maryland

I. INTRODUCTION 117

II. PREPARATION OF SPLEEN POLYRIBOSOMES 119

III. PREPARATION OF MEMBRANE-BOUND AND FREE RIBOSOMES . . 120

IV. CELL-FREE PROTEIN SYNTHESIS 122

V. BINDING OF RADIOACTIVE POLY U TO RIBOSOMES 125

REFERENCES 126

I. INTRODUCTION

The spleen is a highly specialized organ that subserves both hematopoietic and immunological functions. A wide variety of cell types can be found in the spleen, including both fixed tissue cells and wandering cells that enter and leave through the systemic circulation or the lymphatics. Thus at any one time a large percentage of the spleen cells are transient passengers.

The spleen is of great interest to immunologists because it synthesizes large amounts of antibody and contains a number of different cells involved in antigen handling and in the initiation of immune responses. At least three functionally distinct populations of spleen cells interact in the recognition and response to antigen: (1) macrophages, which are phagocytic cells that process antigen and produce an antigen—RNA complex whose exact role in immunity is under study, (2) thymus-derived antigen reactive cells, which appear to regulate the amplitude of the immune response and may serve to concentrate antigen, and (3) bone marrow-derived antibody-producing cells, which differentiate during the immune response and secrete immunoglobulin molecules of defined antibody specificity.

Cells that make antibody may resemble either lymphocytes or plasma cells when examined by light microscopy. By electron microscopy it can be seen that both cell types have a highly developed endoplasmic reticulum and contain large numbers of membrane-associated polyribosomes. Many free polyribosomes and a well-developed Golgi apparatus are also present. The more mature plasma cells contain electron-dense material that may represent antibody within intracytoplasmic channels. They have the characteristics of cells highly specialized for secretion.

It is estimated that between 1 and 5% of spleen cells will make antibody when provoked by a maximal antigenic stimulus. It is important to realize, therefore, that the vast majority of proteins made in the spleen are not γ-globulins (antibodies) and may be proteins destined for intracellular functions rather than secretion. Nevertheless, much interest in protein synthesis by the spleen has developed from attempts to study antibody formation and differentiation in this organ following the administration of antigen and adjuvant.

II. PREPARATION OF SPLEEN POLYRIBOSOMES

Rabbit, mouse, and rat spleens are the organs most often studied. It is important that all procedures be performed at or below 4°C (either in an ice bucket or in a cold room) in order to minimize the hydrolytic action of RNase. The spleen is a rich source of RNase which rapidly hydrolyzes mRNA, causing a breakdown of polyribosomes and an artifactual augmentation of monomeric ribosomes [1].

Male Sprague-Dawley rats (100—200 g) taken from an animal colony (without any special diet beforehand) are killed by decapitation. (Procedures for preparation of liver extracts often specify starvation of animals for several hours or overnight to deplete the glycogen supply of the liver to facilitate centrifugation. This is obviously not necessary with the spleen). Their spleens are removed and immediately immersed in a chilled medium containing 0.25 M sucrose, 0.025 M KCl, and 0.005 M $MgCl_2$ buffered at pH 7.8 with 0.05 M tris. Higher Mg^{2+} concentrations favor artifactual aggregation of ribosomes, while lower

Mg^{2+} concentrations can lead to irreversible dissociation into ribosomal subunits.

Spleens are trimmed, weighed, and homogenized in 4 vol of the above medium with 10 strokes of a loosely fitting Dounce glass homogenizer. The homogenate is filtered through two layers of cheesecloth and centrifuged at 2000 rpm (600 × g) for 5 min to remove unbroken cells, nuclei, and debris. The supernatant is centrifuged at 15,000 rpm (27,000 × g) for 15 min to sediment mitochondria. The resulting supernatant is treated with 1% sodium deoxycholate (DOC) for 20 min. This detergent liberates ribosomes from membranes. The treated supernatant is then layered onto 0.05 M sucrose which in turn sits above a dense layer of 2.0 M sucrose in a centrifuge tube for a Spinco 40 rotor. These sucrose solutions are made up in the same buffer used for extraction. The material is centrifuged at 40,000 rpm for 3 hr. The dense (2.0 M) sucrose permits the passage predominantly of large ribosomal aggregates which are collected as a compact pellet at the end of the centrifugation. These aggregates are resuspended by gentle homogenization. The yield is approximately 0.8 mg/g of spleen [2], with a ratio of A_{260}/A_{280} of 1.8—1.9.

III. PREPARATION OF MEMBRANE-BOUND AND FREE RIBOSOMES

The ribosomes in spleen, as in liver and some other tissues, exist both free in the cytoplasm and associated with membranes. It is thought that membrane-associated ribosomes synthesize proteins destined for export, while free ribosomes produce intracellular proteins [3]. Each ribosome fraction (free and associated) can be prepared free of the other by a modified sucrose gradient method [4].

The initial spleen homogenate is centrifuged for 10 min at 11,500 × g. The resulting supernatant is centrifuged for 90 min at 106,000 × g to prepare a microsome fraction. The microsomal pellet is suspended in homogenizing medium and placed onto a 10—34% linear sucrose gradient (40 ml total volume) which in turn is supported on a 10-ml layer of 68% sucrose. The microsomes are centrifuged for 1 hr at 25,000 rpm in a Spinco SW-25.2 rotor. All sucrose solutions contain 0.025 M KCl, 0.005 M $MgCl_2$, and 0.05 M tris (pH 7.8). The gradient can be collected from the bottom of the tube and can be automatically monitored for absorbance. The membrane-bound fraction collects as an opalescent layer at the interface between the bottom of the gradient and the 68% sucrose, whereas the free ribosomes are distributed in the gradient. The bound and free fractions are pooled separately, diluted to 60 ml with homogenizing medium, and made up to 0.5% of sodium DOC to liberate the ribosomes from membranes. The ribosomes can be collected by centrifugation at 106,000 × g for 2 hr.

The polyribosome profile of each fraction can be determined by ultracentrifugation in a linear 10—34% sucrose gradient (see, for example, Figure 1 of Ref. 2, and Figure 3 of Ref. 3). The free fraction contains many ribosomal aggregates, whereas the bound fraction prepared in this manner consists mainly of single ribosomes. This dissociation of bound polyribosomes is probably a result of the action of RNase, since the inclusion of a 10—25% rat liver supernatant fraction [5] as an RNase inhibitor throughout all procedures preserves polyribosomes in the bound fraction. (For a discussion of the mechanism of action of the rat liver supernatant, see Ref. 3. This supernatant stimulates translation of polyuridylic acid (poly U) and also preserves the structure of polysomes.) The nature of the RNase inhibitor in the rat liver supernatant is not known, and other known RNase inhibitors such as bentonite are without effect. Where hydrolysis by nucleases is a problem, it seems advisable to include the rat liver supernatant in the homogenizing and other media.

IV. CELL-FREE PROTEIN SYNTHESIS

Rat spleen microsomes or polyribosomes stimulate the incorporation of radioactive amino acids into protein in the presence of an energy source and the necessary supernatant factors. The latter are usually supplied by a "pH 5 fraction", which can be prepared from a liver or spleen supernatant in the following manner. The supernatant obtained after centrifugation at 15,000 rpm (27,000 × g) for 15 min is used. Two volumes of a solution containing 0.9 M sucrose, 0.07 M KCl, 4 mmoles/liter $MgCl_2$, and 6 mmoles/liter mercaptoethanol are added to this supernatant. Microsomes are sedimented by centrifugation at 40,000 rpm for 2 hr. The supernatant is adjusted to pH 5 with 1 M acetic acid, allowed to stand at 4°C for 15 min, and centrifuged at 10,000 rpm (12,000 × g) for 10 min. The precipitate is rinsed and suspended in a solution containing 0.35 M sucrose, 0.07 M KCl, 4 mmoles/liter $MgCl_2$, 6 mmoles/liter mercaptoethanol, and 0.05 M tris (pH 7.8). The volume is adjusted to a final protein concentration of 15—20 mg/ml. The activity of this fraction is stable for 2 weeks at −20°C.

Cell-free amino acid incorporation (see also Chapters 1-3) is performed in conical test tubes incubated for 60 min at 37°C. The complete incubation medium contains, in 0.5 ml: 215 mM sucrose, 30 mM tris buffer (pH 7.4), 45 mM KCl, 9.5 mM $MgCl_2$, 4 mM mercaptoethanol, 2 mM ATP, 0.5 mM GTP, 9 mM phosphoenolpyruvate, 35 μg of crystalline phosphoenolpyruvate kinase, 0.08 mM [^{14}C]leucine or [^{14}C]phenylalanine, 0.4 mg of ribosomal protein, and 1.2 mg of pH 5 fraction protein. The polyribosomes are used within 48 hr of preparation. At the end of the 60-min incubation, protein is precipitated by the addition of 5 ml of 5% trichloracetic acid (TCA). The samples are kept at 4°C for 30 min or longer, heated to 90°C for 30 min, washed twice with cold 5% TCA and once with ethanol, and dissolved in 1 ml of hydroxide of Hyamine (Packard Instrument Company, La Grange, Illinois). The samples can then be assayed for radioactivity in a liquid scintillation counter.

In the above procedure protein synthesis is directed by naturally occurring mRNA present on the polyribosomes. For some purposes it may be desirable to study protein synthesis directed by a synthetic messenger such as poly U, which promotes the incorporation of phenylalanine into polyphenylalanine. Under these circumstances incorporation is enhanced by the prior removal of the natural mRNA. This is accomplished by a preliminary incubation in complete medium (without isotope) for 30 min, after which fresh medium (containing [^{14}C]phenylalanine) is added and 100 μg of poly U are introduced into half the tubes. After incubation for 60 min at 37°C, 5% TCA is added and radioactivity determined. The net stimulation by the synthetic messenger represents the difference between tubes containing poly U and those lacking it. Other synthetic mRNAs (coding for various amino acids) can be studied in a like manner.

The incorporation of phenylalanine stimulated by poly U is greater if the pH 5 fraction is prepared from rat liver rather than spleen. Incorporation is further enhanced by preparation of ribosomes in the presence of the already recommended rat liver supernatant, by collecting the ribosomes without sedimentation through 2.0 M sucrose, and by washing the ribosomes in 0.5 M NH_4Cl. These effects are more pronounced on the bound ribosomes which respond poorly to poly U unless these additional procedures are employed. The incorporation of amino acids stimulated by natural mRNA is also augmented slightly by these additional steps.

V. BINDING OF RADIOACTIVE POLY U TO RIBOSOMES

In studies comparing the functional properties of different ribosomes, it may be helpful to determine the amount of synthetic mRNA that can associate with those ribosomes. To do this, ribosomes are incubated at 37°C in a complete amino acid-incorporating medium as described above, except that the radioactive amino acid is omitted and the Mg^{2+} concentration is 24 mM. After 30 min 25—100 μg of [^{14}C]poly U are added, and the tubes are placed in an ice bucket for 5 min. The ribosomes are separated from the incubation medium by centrifugation at 106,000 × g for 90 min. The ribosomal pellet is rinsed once with incubation medium and the rinse combined with the supernatant. Yeast RNA (1 mg) is added to the soluble material and precipitated with 70% ethanol containing 0.2 M NaCl. Any [^{14}C]poly U that was not associated with ribosomes coprecipitates with the carrier yeast RNA. The ribosomal pellet is suspended in ethanol—NaCl. The ribosome and supernatant precipitates are washed three times with ethanol—NaCl, dissolved in 1 ml of hydroxide of Hyamine, and assayed for radioactivity.

REFERENCES

[1] K. H. Stenzel, W. D. Phillips, D. D. Thompson, and A. L. Rubin, Proc. Natl. Acad. Sci. U.S., 51, 636 (1964).

[2] N. Talal, J. Biol. Chem., 241, 2067.

[3] P. Siekevitz and G. E. Palade, J. Biophys. Biochem. Cytol., 7, 619 (1960).

[4] N. Talal and H. B. Kaltreider, J. Biol. Chem., 243, 6504 (1968).

[5] G. Blobel and V. R. Potter, J. Mol. Biol., 28, 539 (1967).

Chapter 5

PROTEIN SYNTHESIS IN EXTRACTS OF WHEAT EMBRYO

Abraham Marcus

The Institute for Cancer Research
Fox Chase
Philadelphia, Pennsylvania

I. POLY U-DEPENDENT INCORPORATION OF PHENYLALANINE . . . 128

A. Preparation of Deoxycholate (DOC)-Washed Ribosomes and Germ Supernatant 129

B. Preparation of Phe-tRNA 130

C. Assay of Phe-tRNA Polymerization 131

D. Preparation of Chain Elongation Factors (T1 and T2) 132

II. TOBACCO MOSAIC VIRUS (TMV) RNA-DEPENDENT INCORPORATION OF AMINOACYL-tRNA INTO POLYPEPTIDES 135

A. Assays 136

B. Preparation of Ribosomes and Supernatant (S-100) 138

C. Separation of Initiation Factors C and D 139

D. Purification of Factor C 140

E. Purification of Factor D 140

F. Preparation of Auxiliary Elongation Factors T1(E) and T2(E) 141

G. Ribosome Binding of Met-tRNA 143

REFERENCES 144

Protein biosynthesis in wheat embryo has been studied in detail with two messenger-dependent amino acid-incorporation systems. The first of these uses polyuridylic acid (poly U) to catalyze the incorporation of phenylalanine from phenylalanyl-tRNA (Phe-tRNA) [1] and can be prepared either from commercial wheat germ or from viable wheat embryos [2]. The second system [3,4] uses a plant viral RNA to catalyze aminoacyl transfer to peptide from aminoacyl-tRNA. This system has thus far been prepared from viable embryos. Another potentially useful reaction for the study of wheat-embryo protein synthesis is that catalyzed by polyribosomes prepared from viable embryos allowed to germinate for 4—6 hr [5]. This reaction has thus far been used primarily as a model of an amino acid-incorporating system that does not require a messenger—ribosome attachment reaction. The following sections describe the preparation and some properties of the first two of these systems. The preparation of the polyribosome system is detailed in Ref. 5.

I. POLY U-DEPENDENT INCORPORATION OF PHENYLALANINE

This reaction system measures phenylalanine transfer to peptide from Phe-tRNA in a reaction that requires ribosomes and two chain-elongation (transfer) factors. In contrast to the incorporation reaction with natural messenger (see Section II), the poly-U reaction does not require initiation factors. It thus serves both as a system for analyzing the details of peptide chain elongation (without the complication of the initiation reaction) as well as a system for assaying the elongation factors. The latter point is particularly useful in ascertaining the extent of contamination of initiation factors by elongation factors.

A. Preparation of Deoxycholate (DOC)-Washed Ribosomes and Germ Supernatant (GS-100)

Commercial wheat germ (30 g) is blended at low speed for 50 sec (five blendings, each of 10-sec duration) in 240 ml of 1 mM $MgAc_2$—2 mM $CaCl_2$—50 mM KCl—6 mM $KHCO_3$. After centrifugation for 10 min at 24,000 × g, 0.01 vol of 0.1 M $MgAc_2$ and 0.025 vol of 1 M tris—acetate (pH 7.6) are added, and the suspension is recentrifuged; first for 10 min at 24,000 × g, and then for 2.5 hr at 78,000 × g in a Spinco S-30 rotor. The supernatant fluid is removed, made 10^{-3} M in dithiothreitol (DTT), and stored at -15°C.

The ribosomal pellet from each Spinco tube is washed twice by suspending in 6 ml of suspension medium [2 mM tris—acetate (pH 7.3), 20 mM KCl, 1 mM $MgAc_2$, 3 mM mercaptoethanol] containing 1% DOC, adding another 4 ml of suspension medium without DOC, and contrifuging for 45 min at 165,000 in a Spinco Ti-50 rotor. The final pellets are suspended in 3 ml of suspension medium containing 20% glycerol and clarified by centrifuging for 10 min at 17,000 × g.

Comments. The DOC treatment is particularly effective in reducing the amount of transfer factors contaminating the ribosomes. The tubes containing the final pellet must, however, be wiped thoroughly to remove adhering liquid since DOC, in appreciable concentration, inhibits amino acid incorporation. An alternative method of removing transfer factors (as yet untested in the wheat embryo system) employs a high-salt wash [6,7], while a third procedure, involving treatment with N-ethylmaleimide [8,9], is apparently of general utility in inactivating the T2 (translocase) ribosomal contaminant.

B. Preparation of Phe-tRNA

A charging incubation contains, per milliliter: [^{14}C]-phenylalanine (1.25 μCi of 300 Ci/mole, 10 mM $MgAc_2$, 43 mM tris—acetate (pH 8), 30 mM KCl, 1.3 mM ATP, 10 mM creatine phosphate, 53 μg of creatine phosphate kinase, 6 mM DTT, 1 mg of wheat germ tRNA [10], and 0.06 ml of S-100 dialyzed for 2 hr against 2 mM tris—acetate (pH 7.6) —5 mM mercaptoethanol —50 mM KCl. [The S-100 is freshly prepared either from viable wheat embryo (see Section II) or from wheat germ (1.1 g ground in a mortar in 8.8 ml of 2 mM $MgAc_2$—2 mM $CaCl_2$—90 mM KCl—10 mM $KHCO_3$—10 mM tris—acetate (pH 7.6), and centrifuged 10 min at 24,000 × g and 60 min at 165,000 × g)].

After 30 min at 20°C, the mixture is deproteinized by three successive extractions with an equal volume of phenol saturated with 0.01 M EDTA (pH 7.6), KAc (pH 5) is added to a final concentration of 0.1 M, and the solution is precipitated with 2.5 vol of ethanol. After overnight storage at -20°C, the precipitate is dissolved in water and dialyzed for several hours against two changes of water. The product contains approximately 350,000 cpm per milligram of tRNA.

Nucleotide-free Phe-tRNA is prepared as follows. The ethanol precipitate obtained from the charging incubation is dissolved in 2 ml of water, and the solution is passed through a 4.5 × 0.65 cm column of DEAE-cellulose equilibrated with 0.3 M KCl—0.01 M KAc (pH 5). The column is then washed with 6 ml of 0.3 M KCl—0.01 M KAc (pH 5) and eluted with 1 M KCl—0.01 M KAc (pH 5). The first 0.8 ml of the 1 M KCl eluant is discarded, and the ensuing 4 ml are collected. The latter solution, containing 60—70% of the input radioactivity, is essentially nucleotide-free and can be concentrated by repeating the ethanol precipitation described above. Alternatively, it can be employed directly after it is dialyzed against several changes of water.

C. Assay of Phe-tRNA Polymerization

A 0.4-ml reaction mixture containing 50 mM tris—acetate (pH 8.0), 4 mM phosphoenolpyruvate, 15 μg of pyruvate kinase, 2.2 mM DTT, 6.5 mM $MgAc_2$, 70 mM KCl, 50 μM GTP, 10 μg of poly U, ribosomes (150 μg of RNA), [^{14}C] Phe-tRNA (15 pmoles, 6000 cpm), and transfer factors is incubated for 10 min at 30°C. The reaction is stopped by the addition of 0.2 ml of 10% trichloroacetic acid (TCA). Bovine serum albumin (100 μg) and 4 ml of 5% TCA are then added, and the suspension is kept on ice for 5 min. The insoluble material is then collected by centrifugation, resuspended in 5% TCA, heated for 15 min at 90°C, cooled on ice for 8 min, and collected by suction filtration on glass-fiber discs (Whatman GF/C). After they are washed with 5% TCA, the discs are dried thoroughly under an infrared lamp and counted in vials containing 10 ml of toluene fluor (400 mg dimethyl POPOP and 500 mg of PPO/liter of toluene).

D. Preparation of Chain Elongation Factors (T1 and T2)

The procedures for partial purification of these factors from commercial wheat germ have been described in detail [1]. The basic assay for each of the factors is the polymerization of $[^{14}C]$ Phe-tRNA (Section I,C) in the presence of a saturating quantity of the second factor. In addition, the activity of each of the factors can be monitored by its partial reaction.

Factor T1 catalyzes the binding of Phe-tRNA to ribosomes [11]. This reaction is assayed under incubation conditions identical to those of the polymerization assay, except that Mg^{2+} concentration is 4.5 mM and phosphoenolpyruvate and pyruvate kinase are omitted. After 10 min at 30°C, 4 ml of diluting buffer [10 mM tris—acetate (pH 7.5), 70 mM KCl, 10 mM $MgAc_2$] are added and the mixture is passed under vacuum through a washed (with diluting buffer) cellulose nitrate filter. The filter is then washed twice with diluting buffer, dried, and counted.

Comments. Under the assay conditions described, nonenzymatic binding is minimal and the GTP requirement is absolute (nucleotide-free Phe-tRNA must be used for these experiments). GMP-PCP (β,γ-methylene-guanosine triphosphate) is about 30% as effective (at saturation) as GTP. If the Mg^{2+} concentration is raised to 6.5 mM, nonenzymatic binding is still negligible. However, considerable GTP-independent enzymatic binding is observed, and the GMP-PCP reaction is about 60% as effective as the GTP reaction.

When [^{14}C]Phe-tRNA is first enzymatically bound to ribosomes in the presence of GMP-PCP and is subsequently converted to polyphenylalanine by a second incubation in the presence of GTP, T2, and a large excess of [^{3}H]Phe-tRNA, the product contains more than 60% of its ^{14}C and less than 15% of its ^{3}H at the N terminus. It therefore appears that the GMP-PCP-catalyzed binding reaction is an authentic component of the polymerization reaction. An ensuing reaction, presumably involving GTP, may shift the GMP-PCP-bound Phe-tRNA to a position where it is reactive (forming a new peptide bond) with peptidyl-tRNA present on the ribosome at the donor or peptidyl site. Such a reaction has, however, not as yet been reported.

Factor T2 catalyzes the translocation of ribosome-bound Phe-tRNA; its activity can be assayed by the reactivity of bound Phe-tRNA with puromycin (see also Chapters 1, 3, and 9). Experimentally, the assay consists of an initial nonenzymic binding reaction, followed by a T2-catalyzed translocation and a subsequent reaction with puromycin. The components of the initial nonenzymic binding reaction are identical with those of the polymerization assay except that GTP and soluble factors are omitted; the volume is 0.10 ml, the Mg^{2+} concentration is 20 mM, and the incubation is for 6 min at 20°C. Thereafter GTP (7×10^{-5} M) and translocase (T2) are added; the volume is brought to 0.30 ml (with KCl maintained at 70 mM), and the incubation is continued for 10 min at 30°C. GMP-PCP (7×10^{-4} M) and puromycin (6×10^{-4} M) are then added to a volume of 0.35 ml. The incubation is continued for 30 min at 30°C, at which time 1 ml of 2 M NH_4HCO_3 and 3 ml of ethyl acetate are added, and the radioactivity extracted into the ethyl acetate layer (phenylalanyl-puromycin derivatives) is determined [12].

Comments. The key feature of the translocase assay is that the Mg^{2+} concentration must be lowered to 5—8 mM during the second incubation. This is in clear contrast to the bacterial reaction [13].

The nature of the puromycin product has not as yet been determined. If, however, additional Phe-tRNA is added during the translocase step (the second incubation), considerably more of the Phe-tRNA bound nonenzymically in the first incubation, as well as approximately equimolar amounts of the Phe-tRNA added in the second incubation, are found in the puromycin product. Since when Phe-tRNA is added only in the second incubation (and omitted from the first incubation) there is no puromycin product, it appears that the normal product of the two-step reaction may be diphenylalanyl-puromycin. If this is true, then nonenzymic binding of Phe-tRNA in the wheat embryo would require an unusual (low Mg^{2+}) type of binding, with such binding occurring only when the peptidyl site is occupied.

II. TOBACCO MOSAIC VIRUS RNA-DEPENDENT INCORPORATION OF AMINOACYL-tRNA INTO POLYPEPTIDES

This overall reaction utilizes at least two initiation factors (C and D) and two elongation factors (T1 and T2), all of which are present in the supernatant (S-100) fraction. Each of the four factors has been prepared free of the other three components, with their relative affinity for DEAE-cellulose being the primary basis of separation. The order of decreasing retention is factor D, T2, factor C, and T1. This section describes the procedures used in the isolation and partial purification of initiation factors C and D. In addition, auxiliary methods are described for obtaining highly active preparations of elongation factors from column eluants obtained as byproducts of the purification of the initiation factors.

A. Assays

The primary assay utilized to purify the initiation factors takes advantage of the specificity of aurintricarboxylic acid (ATA) in inhibiting the initiation reaction [14,15]. An initial incubation is carried out under conditions permitting the formation of ribosome—messenger initiation complexes. The extent of initiation that has occurred in this incubation is then monitored by adding ATA to block further initiation and ascertaining the capacity of the system for aminoacyl transfer to peptide in the presence of an excess of elongation factors. The preliminary incubation contains, in a volume of 0.35 ml: 10 μmoles tris—acetate (pH 8), 0.4 μmole ATP, 3.2 μmoles creatine phosphate, 16 μg of creatine phosphate kinase, 0.01 μmole GTP, 10 μg of TMV RNA [16], 0.9 μmole DTT, ribosomes (once washed, 200 μg of RNA), 0.84 μmole $MgAc_2$, 17.5 μmoles KCl, and soluble factors; it is carried out for 6 min at 30°C. Thereafter 0.03 ml of S-100 (undialyzed), 0.6 μmole $MgAc_2$, 12 nmoles ATA (final concentration 3×10^{-5} M), and [^{14}C]aminoacyl-tRNA (17,000 cpm) are added, to a final volume of 0.41 ml. After the tubes are incubated for another 9 min at 30°C, the radioactive material insoluble in hot TCA is determined (see Section I,C).

[^{14}C]Aminoacyl tRNA is prepared by a procedure identical to that described for Phe-tRNA (Section I,B) except that the following eight [^{14}C]amino acids are added, at a concentration of 1.25 μCi/ml of charging incubation: leucine, serine, lysine, glutamic acid, threonine, valine, proline, and phenylalanine. In addition, the tris—acetate concentration is raised to 83 mM to compensate for the additional HCl added with the radioactive amino acids. The final product contains 10^6 cpm/mg of tRNA.

An alternative direct assay for screening the incorporating capacity of either ribosomes or S-100 uses a single incubation with components as described above for the preincubation except for the following changes: final volume 0.40 ml; K^+ concentration, 44 mM; Mg^{2+} concentration, 3.66 mM; 12 μg of tRNA; and 0.13 μCi of [^{14}C]leucine. After 30 min at 30°C, a standard system (ribosomes, 200 μg of RNA; 0.08 ml of S-100 dialyzed for 2 hr against 1 mM tris (pH 7.3)—50 mM KCl—4 mM mercaptoethanol) results in the incorporation of 20—30,000 cpm.

Transfer factors T1 and T2 are assayed with the poly U system, as described in Section I,C and D.

Comments. The assay systems described are essentially completely dependent upon a source of exogenous messenger. This may be supplied by one of several plant viral RNAs, TMV RNA, turnip yellow mosaic virus (TYMV) RNA, satellite tobacco necrosis virus (STNV) RNA, or by an endogenous messenger fraction from wheat embryo [17]. MS2 RNA is, however, inactive.

In this latter regard, the wheat embryo system clearly differs from its bacterial counterpart. It may also be noted that one other major difference exists, namely, the absolute requirement for ATP in the wheat embryo system [4].

B. Preparation of Ribosomes and Supernatant (S-100)

Viable wheat embryos (1.1 g) are ground with a cooled mortar and pestle in 11 ml of 1 mM $MgAc_2$—2mM $CaCl_2$—90 mM KCl—6 mM $KHCO_3$. After centrifugation for 10 min at 24,000 X g, 0.1 vol of both 0.1 M $MgAc_2$ and 1 M tris—acetate (pH 7.6) is added and the suspension is centrifuged for 10 min at 24,000 X g and then for 60 min at 150,000 X g. The top three-fourths of the supernatant fluid is taken as S-100 and stored at -15°C.

The ribosomal pellet is suspended in 10 ml of suspension medium (2 mM tris (pH 7.4)—20 mM KCl—1 mM $MgAc_2$—3 mM 2-mercaptoethanol) and centrifuged for 45 min at 150,000 X g. The pellet obtained is suspended in 2 ml of suspension medium containing 20% glycerol and clarified by centrifugation for 10 min at 17,000 X g. Such ribosomes are referred to as once-washed ribosomes. In an alternative procedure the pellet obtained from the first wash is suspended in 3 ml of suspension medium and layered over 7 ml of suspension medium containing 25% sucrose. The pellet obtained after centrifugation for 135 min at 150,000 X g is processed as above. This preparation is referred to as twice-washed ribosomes.

Comments. The pH of the initial homogenization should be 6.3—6.7 (as checked after the first centrifugation at 24,000 X g). If the pH is lower than 6.3, preparations of reduced activity are obtained. A higher pH results in contamination of ribosomes by endogenous messenger [17], which otherwise is normally sedimented during the first centrifugation.

The majority of incorporation studies have been carried out with once-washed ribosomes. The second wash (through the sucrose cushion) reduces the tRNA content of the ribosomes, although some tRNA contamination still remains. Twice-washed ribosomes also show a much greater selectivity for the initiating species of Met-tRNA in the binding assay (see Section II,G). DOC-washed ribosomes (as in Section I,A) are completely inactive.

C. Separation of Initiation Factors C and D

Seven milliliters of S-100 are chromatographed on a 5 X 0.7 cm DEAE-cellulose (Whatman DE23) column equilibrated with medium II (1 mM tris—acetate (pH 7.0)—2 mM $MgAc_2$—4 mM mercaptoethanol) containing 0.1 M KCl. Elution of the column is continued with medium II—0.1 M KCl. A void volume of 2 ml is discarded; 7 ml are collected as fraction C, and a subsequent 5-ml wash fraction is discarded. The eluant is changed to medium II—0.3 M KCl; 1 ml is discarded and 3.8 ml are collected as fraction D. Both of these fractions are relatively stable when kept at -20°C; they are used as reagents (0.06 ml of fraction C and 0.05 ml of fraction D) in the initiation assay (Section II,A) to follow the purification of the other component.

D. Purification of Factor C

(1) A 5.5-ml portion of fraction C is dialyzed for 45 min against medium III (1 mM tris—acetate (pH 7.6)—1 mM $MgAc_2$—4 mM mercaptoethanol—0.1 mM EDTA) and immediately chromatographed on a 9 X 0.9 cm DEAE-cellulose column equilibrated with medium III. Elution of the column is continued with medium III—30 mM KCl, and a total volume of 10 ml is discarded. The eluant is changed to medium III—0.15 M KCl; 2 ml are discarded and 6 ml are collected. (2) Four milliliters of the product of step (1) are dialyzed for 30 min against medium II—0.1 mM EDTA—30 mM KCl and immediately applied to a 4.5 X 0.5 cm cellulose P11 column that has been washed with medium II—0.1 mM EDTA—50 mM KCl. Elution is continued with medium II—0.1 mM EDTA—50 mM KCl, and a total volume of 9.2 ml is discarded. The eluant is changed to a solution containing 20 mM tris—acetate (pH 7.4), 1 mM EDTA, 10% glycerol, 0.5 M KCl, and 0.5 mM DTT, and 3.5 ml are collected as C final.

E. Purification of Factor D

(1) The initial procedure for obtaining fraction D is modified (see Section II,C) as follows. After obtaining fraction C the eluant is changed to medium II—0.15 M KCl, and

a total of 14 ml is collected and discarded. The eluant is changed to medium II—0.3 M KCl; 1 ml is discarded and 3.8 ml are collected. (2) Three milliliters of the product of step (1) are dialyzed for 60 min against 1 mM tris (pH 7.6)—1 mM $MgAc_2$—4 mM mercaptoethanol—0.15 M KCl and chromatographed on a 5 X 0.7 cm DEAE-cellulose column equilibrated with the dialysis solution. Elution of the column is continued with the same solution, and a total volume of 10 ml is discarded. The eluant is changed to medium II—0.3 M KCl; 1 ml is discarded and 3.5 ml are collected as D final.

F. Preparation of Auxiliary Elongation Factors T1(E) and T2(E)

T1(E)

The 12 ml (10 ml + 2 ml) normally discarded in step (1) of Section II,D (purification of factor C) are collected, and 3.8 g of solid $(NH_4)_2SO_4$ are added (45% saturation). The pellet obtained after centrifugation is discarded and an additional 0.85 g of $(NH_4)_2SO_4$ (55% saturation) is added. After overnight storage at 4°C, the suspension is centrifuged and the pellet is dissolved in 0.7 ml of medium II—50 mM KCl, dialyzed for 2 hr against 1 mM tris—acetate (pH 7.3)—50 mM KCl—4 mM mercaptoethanol, and diluted with an equal volume of medium II—50 mM KCl—60% glycerol.

T2(E)

The 15 ml (14 ml + 1 ml) normally discarded in step (1) of Section II,E (purification of factor D) are collected, and 3.71 g of solid $(NH_4)_2SO_4$ are added (35% saturation). The pellet obtained after centrifugation is discarded, and an additional 1.59 g $(NH_4)_2SO_4$ are added (50% saturation). After overnight storage at 4°C, the pellet is collected by centrifugation and processed as described for T1(E).

Comments. The incorporation reaction is quite sensitive to the concentration of monovalent cations [18]. Since there is considerable K^+ in the solutions of the various factors, this must be corrected for in the assay. In most cases the factor solutions are dialyzed for 1.5—2 hr against 1 mM tris—acetate (pH 7.3)—50 mM KCl—4 mM mercaptoethanol, and their K^+ concentration is taken to be 50 mM.

The purification procedures for initiation factors C and D result in approximately 10-fold purification of each of the factors, as well as the complete removal of elongation factors T1 and T2. The two auxiliary elongation factors T1(E) and T2(E) are also free of each other and of factors C and D. These separations make it possible to carry out total reconstitution experiments. With regard to stability, both elongation factors can be stored at -20°C for several weeks with little loss of activity. The purified initiation factors are, however, considerably less stable.

G. Ribosome Binding of Met-tRNA

A useful adjunct in demonstrating initiation factor activity is a ribosome-binding reaction specific for the initiating species of Met-tRNA [19]. The two species of Met-tRNA ($tRNA_i^{Met}$, the initiating species and $tRNA_m^{Met}$, the internal species) are obtained by chromatography on benzoylated DEAE-cellulose [20,21] and may be charged as described for aminoacyl-tRNA (see Section II,A). The components of the binding assay are identical to those of the preliminary incubation (see Section II,A) with the following changes: creatine phosphate and creatine phosphate kinase are omitted; GTP concentration, 50 μM; Mg^{2+} concentration, 1.3 mM; and twice-washed ribosomes are used. The reaction is started by the addition of Met-$tRNA_i$ (3000 cpm) and is incubated for 5 min at 20°C. Ribosome-bound radioactivity is then determined as in Section I,D.

Comments. The key feature of the ribosome binding assay is that the Mg^{2+} concentration is reduced to 1.3 mM. At higher concentrations of Mg^{2+}, codon-dependent binding catalyzed by elongation factor T1 becomes appreciable. Under the assay conditions described, some Met-tRNA binding is observed when either factor C or factor D is added alone. However, when both factors are present together, binding is considerably augmented relative to their individual activities. Factors T1 and T2 are completely inactive.

ACKNOWLEDGMENTS

This work was supported by grant GB-23041 from the National Science Foundation, U.S.P.H.S. grants CA-06927 and RR-05539 from the National Institutes of Health, and by an appropriation from the Commonwealth of Pennsylvania.

REFERENCES

[1] A. B. Legocki and A. Marcus, J. Biol. Chem., 245, 2814 (1970).

[2] F. B. Johnston and H. Stern, Nature, 179, 160 (1957).

[3] A. Marcus, B. Luginbill and J. Feeley, Proc. Nat. Acad. Sci. USA, 59, 1243 (1968).

[4] A. Marcus, J. Biol. Chem., 245, 955, 962 (1970).

[5] D. P. Weeks and A. Marcus, Plant Physiol., 44, 1291 (1969).

[6] A. App, Plant Physiol., 44, 1132 (1969).

[7] J. Siler and K. Moldave, Biochim. Biophys. Acta, 195, 123 (1969).

[8] R. D. Mosteller, W. J. Culp,and B. Hardesty, Proc. Nat. Acad. Sci. USA, 57, 1817 (1967).

[9] W. L. McKeehan and B. Hardesty, J. Biol. Chem., 244, 4330 (1969).

[10] G. Zubay, J. Mol. Biol., 4, 347 (1962).

[11] M. Nirenberg and P. Leder, Science, 145, 1399 (1964).

[12] P. Leder and H. Bursztyn, Biochem. Biophys. Res. Commun., 25, 233 (1966).

[13] N. Brot, R. Ertel, and H. Weissbach, Biochem. Biophys. Res. Commun., 31, 563 (1968).

[14] A. P. Grollman and M. L. Stewart, Proc. Nat. Acad. Sci. USA, 61, 719 (1968).

[15] A. Marcus, J. D. Bewley, and D. P. Weeks, Science, 167, 1735 (1970).

[16] H. Fraenkel-Conrat, in (G. L. Cantoni and D. R. Davies, eds.) Procedures in Nucleic Acid Research, Harper, New York, 1966, p. 480.

[17] D. P. Weeks and A. Marcus, Biochim. Biophys. Acta, in press (1971).

[18] J. D. Bewley and A. Marcus, Phytochemistry, 9, 1031 (1970).

[19] A. Marcus, D. P. Weeks, J. P. Leis, and E. B. Keller, Proc. Nat. Acad. Sci. USA, 67, 1681 (1970).

[20] J. P. Leis and E. B. Keller, Biochem. Biophys. Res. Commun., 40, 416 (1970).

[21] H. Tarrago, O. Manasterio, and J. E. Allende, Biochem. Biophys. Res. Commun., 41, 765 (1970).

Chapter 6

CEREBRAL PROTEIN-SYNTHESIZING SYSTEMS

Claire E. Zomzely-Neurath

Department of Biochemistry
Roche Institute of Molecular Biology
Nutley, New Jersey

I. INTRODUCTION 147

II. PREPARATION OF CEREBRAL MICROSOMES, RIBOSOMES, AND pH 5 ENZYMES 149

A. Microsomes 149

B. Ribosomes. 153

C. pH 5 Enzymes 155

III. PREPARATION OF TOTAL AND FREE POLYRIBOSOMES. 159

A. Total Polyribosomes. 159

B. Free Polyribosomes 166

IV. CELL-FREE AMINO ACID-INCORPORATING SYSTEM. 168

V. PREPARATION OF RNA WITH PROPERTIES OF mRNA 176

VI. PREPARATION OF "STRIPPED" RIBOSOMES. 183

REFERENCES 184

I. INTRODUCTION

Cell-free systems derived from brain have been found to be highly active in protein synthesis. Ribosomal and polyribosomal preparations, characterized to a greater or lesser extent, have been described by several investigators [1-9]. The most comprehensive studies have generally used mice or rats as the source of cerebral tissue. Since the eventual goal of most scientists working in the area of cerebral protein synthesis is to be able to relate the effects of a changing functional environment to specific protein synthetic pathways, the use of small animals is almost mandatory. The investigator is able to control many variables such as age, sex, state of health, time and method of killing, interval between death and final preparation, as well as uniformity of colony and environment. Such factors are virtually impossible to maintain consistently when tissue is obtained from large animals slaughtered commercially. Furthermore, investigations of factors affecting, for example, normal development of the brain, effects of drugs, learning, and so on, are facilitated by the use of an animal that has a relatively short life span such as the rat or mouse.

Methods for preparation of the various components of cerebral cell-free protein-synthesizing systems are based on procedures originally developed for hepatic tissue. However, certain modifications in methodology are necessary. These are related to the complexity of the cells within the brain, as compared with the relative uniformity of hepatic cells, and also to apparently intrinsic properties of the cerebral protein-synthesizing system. These intrinsic properties have not completely been defined at this time, nor is their existence universally accepted. However, it is reasonable to believe that the protein-biosynthetic machinery of the brain may possess unique properties which permit or facilitate the synthesis of brain-specific proteins involved in the complex functions of this highly differentiated tissue.

There are several levels of refinement in preparative techniques for the isolation of the components for cell-free protein-synthesizing systems. The choice of method depends, of course, on the particular research program. Although most of the recent studies in the area of cerebral protein synthesis use relatively purified components, a fairly crude system is sometimes sufficient for evaluation of the feasibility of a particular hypothesis. Therefore, the preparative methods that follow are presented in terms of increasing degrees of methodological refinement.

II. PREPARATION OF CEREBRAL MICROSOMES, RIBOSOMES, AND pH 5 ENZYMES

A. Microsomes

In bacteria, yeast, and reticulocytes, most of the ribosomes occur free in the cytoplasm and can be isolated by simple centrifugal procedures. By contrast, in animal cells, particularly those that synthesize proteins for export such as liver and pancreas, most of the cytoplasmic ribosomes are attached to membranes of the endoplasmic reticulum. When the cells are homogenized in sucrose buffered at pH 7.0-7.6, the membranes are broken but form spherical particles with the ribosomes still attached to their outer surfaces. These particles make up the microsomal fraction. In brain, however, particularly when the cortex is considered, neuronal cells appear to possess a higher proportion of free cytoplasmic ribosomes to reticulum-bound ribosomes [10-13]. Although the glial cells surrounding the neurons, as well as those present in "white matter" of brain, contain a higher proportion of bound ribosomes, an appreciable amount of free cytoplasmic ribosomes is also present. Therefore the microsome fraction from brain contains different proportions of free and bound ribosomes depending on whether whole brain or cortex is used for preparation of the microsomes.

After the animals are killed, the brains are quickly removed and placed on filter paper moistened with the preparative medium in ice-cold petri dishes on an ice bath. If only certain regions of the brain are to be used, the dissection should be carried out with the brains maintained cold on the ice bath. The preparative medium should always be kept in a refrigerator or a cold room at 4°C. Homogenization, as well as all further steps, are carried out in a cold room. The preparative medium is composed of 0.25 M sucrose, 25 mmoles/

liter KCl, 4 mmoles/liter $MgCl_2$, and 0.05 M tris-HCl buffer (pH 7.4).

In a cold room the tissue is transferred to a 50-ml beaker in an ice bath, and 1 ml of cold sucrose-tris buffer is added. The tissue is then lightly minced with scissors. This step facilitates the homogenization of the tissue, which is carried out in sucrose-tris buffer in the proportion of 9 ml/g of tissue. Five up-and-down strokes of a Teflon pestle in a glass tube with a smooth inner surface are sufficient for a smooth homogenate. If clearance between the tube and pestle is 0.4 mm, there is no difference in the activity of the microsomal preparation even if as many as seven strokes are used. However, with a tighter clearance, for example, 0.2 mm, three strokes are enough for homogenization. Cell debris, nuclei, and mitochondria are then sedimented by centrifugation of the homogenate at 15,000 X g in a refrigerated centrifuge at 0°C for 15 min. The resulting pellet is firm, and the supernatant can be readily decanted into a beaker in an ice bath. At this point this postmitochondrial supernatant can be centrifuged for 30 min at 34,500 X g (0°C) to obtain a "large" microsome fraction, which contains the heavier microsomal particles.

The large microsome fraction also contains a small amount of free ribosomes. The supernatant from the large microsome fraction can then be transferred to the plastic tubes used for a Spinco 40 fixed-angle rotor. The samples are centrifuged for 2 hr at 40,000 rpm (105,000 X g) at 0°C. The supernatant (postmicrosomal), also called the S-100 fraction, is decanted into a beaker in an ice bath and can be used directly for *in vitro* studies or for preparation of the pH 5 enzymes fraction. The pellets are rinsed three times with sucrose-tris medium and can be either used directly for experiments or stored in the pelleted form at -60°C for at least 4 weeks. These pellets consist of a mixture of free ribosomes and microsomal particles (ribosomes attached to the endoplasmic reticulum). The yield of this fraction is about 1 mg/g of cerebral tissue.

A less time-consuming procedure, which is suitable for most studies, involves the sedimentation of the total microsomal fraction. In this case the postmitochondrial supernatant is centrifuged immediately for 2 hr at 105,000 X g at 0°C in a Spinco 40 rotor without prior sedimentation of the large microsome fraction. This total fraction is less stable on storage than the microsome fraction from which the large microsomes have been separated. However, the activity of this total microsome fraction remains essentially constant for 1-2 weeks at -60°C. Since the microsomes tend to bind an appreciable amount of extraneous proteins such as RNases and proteases during the isolation procedure, the relative instability of the microsomes is attributable, to a large extent, to these bound degradative enzymes. If further purification is desired, the microsomal pellets can be suspended and centrifuged through 1 M sucrose containing the same buffer and salts as the homogenization medium. For sedimentation of the total microsome fraction through 1 M sucrose, 2.5-3 hr of centrifugation at 105,000 X g at 0°C are required. The microsome fraction consists chemically of protein, lipid, and RNA. The cerebral microsome preparations from rat cerebral cortex prepared by the above methods contain 20-22% RNA [14].

B. Ribosomes

The homogenization medium and procedure for preparation of the postmitochondrial supernatant are essentially the same as described for microsomes. In order to obtain the ribosome fraction, the particles bound to the endoplasmic reticulum must be freed from this lipoprotein membranous structure. Sodium deoxycholate (DOC), a bile salt, dissolves these membranes without impairing the physical integrity or amino acid-incorporating activity of the cerebral ribosomes [14-18]. Ribosomes can be isolated directly from the postmitochondrial supernatant fraction after addition of sodium DOC to a final concentration of 0.5%. The detergent-treated postmitochondrial supernatant is then centrifuged at 105,000 X g_{av} at 0°C for 2 hr to pellet the ribosomes. After decantation of the supernatant, the ribosome pellets are rinsed three times with sucrose-tris buffer, resuspended, and centrifuged again at 105,000 X g for 2 hr. These repeated washings remove all traces of DOC, which is inhibitory to the amino acid-incorporating activity of ribosomal preparations [19]. An alternative method consists of treating the pelleted total microsomal fraction with sodium DOC to a final concentration of 0.25%. The rinsing and repelleting of the isolated ribosomes is carried out in the same manner as described above. In this procedure the postmicrosomal supernatant (S-100) can be used as the source of enzymes required for *in vitro* protein synthesis or for preparation of the pH 5 enzymes fraction. When the postmitochondrial supernatant is directly treated with detergent for the isolation of ribosomes, the supernatant fraction remaining after pelleting of the ribosomes cannot be used as a source of enzymes since it contains DOC.

For preparation of ribosomes the ratio of homogenizing medium to tissue can be reduced to 3 ml/g of cerebral tissue since treatment of either the postmitochondrial supernatant or the total microsome pellet reduces binding of extraneous protein. The dilute homogenate (9 ml/g of tissue) used in preparation of microsomes aids in decreasing the nonspecific binding of proteins.

Sodium DOC can be prepared as a concentrated solution in 0.05 M tris-HCl buffer (pH 8.2) at a concentration of 5 or 10%.

It should be prepared shortly before use and kept in an ice bath until ready to be added to either the postmitochondrial supernatant or the microsomal suspension.

Ribosomes prepared from the postmitochondrial supernatant must be carried through the entire procedure before they can be stored. However, if ribosomes are prepared from the total microsome fraction, the procedure can be terminated at the time of pelleting of the microsomes. These can be stored, as stated previously, for 1-2 weeks. Thereafter ribosomes can be prepared from stored microsomes at any time during this interval.

Ribosomes prepared from rat cerebral cortex (Sprague-Dawley male rats, 6 weeks of age) contain 45-50% RNA [18] and are obtained in a yield of 0.6 mg RNA/g of cerebral cortex. These ribosomes are stable for several weeks of storage at -60°C in spite of a slight contamination by RNase [7]. A typical sucrose density gradient pattern of ribosomes isolated from rat cerebral cortex is shown in Figure 1.

C. pH 5 Enzymes

Although the postmicrosomal supernatant can be used for *in vitro* amino acid incorporation, the stability of this fraction on storage at low temperatures is variable. Therefore it is more practical to prepare the pH 5 enzymes fraction since this is stable at -60°C for about 1 year. This stability is especially useful in certain types of studies requiring several

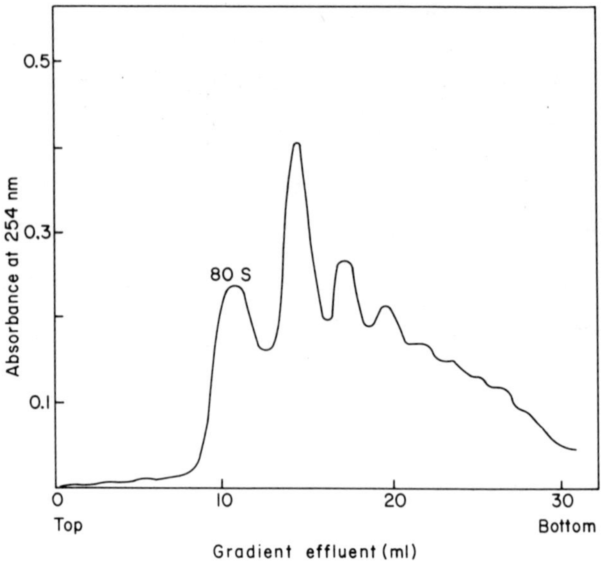

Fig. 1. Sucrose density gradient profile of ribosomes prepared from rat cerebral cortex. A suspension of ribosomes (4.8 A_{260} units) in 2.0 ml of 0.05 M tris-HCl buffer (pH 7.4) containing 4 mmoles/liter $MgCl_2$ and 25 mmoles/liter KCl was layered on a linear sucrose gradient (5-20%) containing the same buffer and salts as the suspension medium. Centrifugation was in a Spinco SW-25 swinging-bucket rotor at 25,000 rpm at 0°C for 3 hr. Absorbance of the effluent collected from the top of the gradient was monitored continuously at 254 nm. An ISCO Model D density gradient fractionator (Instrument Specialties Company) and an ISCO Model UA ultraviolet analyzer with external recorder were used for fractionation of the gradient and measurement of ultraviolet absorbance.

different preparations, such as comparison of different areas of the brain or development of the brain with respect to protein-synthesizing capacities.

The pH 5 fraction is prepared from the postmicrosomal supernatant (S-100) fraction. The pH of this supernatant fraction is adjusted to 4.7 by the slow addition of 1 N acetic acid with constant stirring in an ice bath in a cold room (4°C). The pH of the comparable liver supernatant is usually 5.0-5.2 [20]. However, with cerebral preparations a pH of 4.7 is necessary for precipitation of the desired fraction. If a concentrated homogenate (3 ml of medium per gram of tissue) is used, then the postmicrosomal supernatant should be diluted with sucrose-tris buffer to a final concentration of 9 ml of medium per gram of tissue. If this step is omitted, a less active preparation of pH 5 enzymes is obtained. The postmicrosomal supernatant fraction that has been treated with acetic acid is allowed to remain in the ice bath in the cold room for 1/2 hr. The preparation is then stirred gently, transferred to plastic tubes, and centrifuged at 34,500 X g at 0°C for 15 min. With cerebral preparations adequate packing of the precipitated pH 5 enzymes requires a higher speed of centrifugation than preparations of hepatic origin, which are effectively sedimented at 15,000 X g. The supernatant fluid (pH 5 supernatant) is decanted and the pellets are rinsed three times in cold tris-buffer medium [50 mmoles/liter tris-

HCl (pH 7.4), 25 mmoles/liter KCl, 4 mmoles/liter $MgCl_2$]. The pH 5 precipitate is resuspended gently in the same buffer and centrifuged for 15 min at 5000 X g at 0°C. The supernatant fluid, containing the pH 5 enzymes, is removed and adjusted to pH 7.4 with 0.2 N KOH. The yield of pH 5 fraction per gram of cerebral tissue is 2-4 mg of protein per gram of tissue, depending on the age of the animal and whether whole brain, cortex, or white matter is used as the source of the fraction. Therefore the pH 5 pellets are suspended prior to the low-speed centrifugation in 3 ml of medium per 10 g of tissue. This volume gives a final protein concentration of 8-12 mg of protein per milliliter. The solution of pH 5 enzymes can be stored in aliquots of desired size at -60°C for about 1 year. Actually, the pH 5 fraction can be thawed and refrozen at least three times without any change in activity in the cell-free amino acid-incorporating system.

The postmicrosomal supernatant fraction (also called cell sap) can be passed through a column of Sephadex G-25 to remove free amino acids, nucleotides, and sucrose [21] prior to precipitation of the pH 5 fraction with acetic acid or the direct use of the cell sap for amino acid incorporation studies. This step does not appear to be important in the preparation of pH 5 enzymes from cerebral tissues, in contrast to the situation with hepatic tissue. Sephadex treatment of the hepatic postmicrosomal supernatant fraction apparently removes certain compounds that can be inhibitory in a cell-free amino acid-incorporating system [22].

Preparation of the pH 5 fraction should be carried out as much as possible in a cold room (4°C), and the samples at various stages should always be kept in an ice bath. The procedure cannot be stopped at any point with the intention of leaving the sample overnight at a particular stage. In order to prepare a stable and active preparation, it is necessary to carry out the procedure to the final stage of storage in buffer at -60°C.

III. PREPARATION OF TOTAL AND FREE POLYRIBOSOMES

A. Total Polyribosomes

The concept that polyribosomes are the functional units in protein synthesis in most cells was proposed by several groups of investigators in the early 1960s [23-27]. Their reports contained an imposing body of evidence that formed the basis of this proposal. Since that time, polyribosomes have indeed been found to fulfill the role of protein factory in all living cells that have been investigated. As mentioned previously, in animal cells ribosomes and polyribosomes exist in two forms: essentially free in the cytoplasm, or attached to the endoplasmic reticulum. At any time that an animal is killed, there exists in the cytoplasm a mixture of single ribosomes, small clusters of ribosomes (two or three), and polyribosomes (four or more ribosomes held together by strands of mRNA). Some of these ribonucleoprotein particles (another term for ribosomes) can be seen attached to the endoplasmic reticulum, while others appear as free particles (in electron micrographs). The preparative procedures already described for microsomes and ribosomes involve the isolation of all particle sizes from a single ribosome (80 S particle) to aggregates of several ribosomes (polyribosomes). In the absence of a detergent such as DOC, microsomes are obtained that are a mixture of free ribosomes and those attached to the endoplasmic reticulum. When detergent is used for the liberation of the bound particles, a total ribosome fraction is obtained. These ribosomes are a mixture of all sizes of particles. Furthermore, the preparative procedure, particularly for brain, results in the breakdown of some of the heavier ribosomal aggregates. In order to prepare polyribosomes, a procedure is used in which the heavier ribosomal aggregates (polyribosomes) are separated, to a large extent, from the single ribosomes and the clusters of two or three ribosomes. The method involves the use of a discontinuous gradient of sucrose. The procedure was first reported by Wettstein et al. [27] for the preparation of polyribosomes (also called polysomes) from rat liver. In order to obtain both types of polysomes (bound and free), DOC was used to liberate the bound polysomes. Thus the total polyribosome fraction from brain consists of free and released polysomes, analagous to the preparation obtained by Wettstein et al. [27] from rat liver.

The homogenization medium used for preparation of total polyribosomes from brain consists of 0.25 M sucrose containing 20 mmoles/liter tris-HCl buffer (pH 7.6), 40 mmoles/liter NaCl, 10 mmoles/liter KCl, and 10 mmoles/liter magnesium acetate or $MgCl_2$. Tissues are obtained in the same manner as described for the preparation of microsomes and ribosomes. After the tissue is minced, it is homogenized in sucrose-tris buffer (3 ml of medium per gram of tissue) with three to four up-and-down strokes of a Teflon pestle. Again, all operations are performed in a cold room (4°C), and samples are kept in an ice bath as much as possible. The homogenate is centrifuged at 12,000 X g at 0°C for 10 minutes in order to obtain the postmitochondrial supernatant. A 10% solution of sodium DOC in 0.05 M tris-HCl (pH 8.2) is slowly added, with gentle stirring, to the postmitochondrial supernatant fraction; the final concentration of DOC is 1%. This DOC-treated supernatant is then layered over a 0.5 M sucrose solution that has been previously layered over 2.0 M sucrose, containing the same concentration of tris-HCl buffer and salts as the homogenizing medium. The relative volumes of the three layers are 2:1:1. The discontinuous gradient is centrifuged at 0°C at 105,000 X g for 4 hr. The pellet obtained is rinsed three times with the homogenizing medium and drained; the inner walls of the tubes are wiped with a paper tissue, and the pellets are stored at -60°C. These pellets are stable on storage at this temperature for at least 4 months without any change in sedimentation pattern or amino acid-incorporating activity.

The most important component of the preparative medium, in terms of isolation of polyribosomes with a high proportion of heavy aggregates, is the magnesium concentration. Magnesium, at a concentration of at least 10 mM, is required for a stable, active polyribosome preparation from brain [7,18]. Under ionic conditions (1-5 mM Mg^{2+}) that result in 92-98% of ribosomal components heavier than trimer (three ribosomes) in hepatic polyribosome preparations, the percentage of analogous fractions in cerebral polyribosomes is only 50% or less. By contrast, when 10-12 mM Mg^{2+} is used in the isolation medium, the proportion of components heavier than trimer is increased to 80%. This is not attributable to nonspecific aggregation caused by the high magnesium concentration since these cerebral polyribosomes are dissociated to single ribosomes (80 S) by treatment with pancreatic RNase even more readily than hepatic polyribosomes prepared with 1-5 mM magnesium in the medium. Furthermore, cerebral polyribosomes are just as active as hepatic preparations with respect to amino acid-incorporating activity _in vitro_ [7,18]. Use of a natural inhibitor of RNase isolated from cerebral tissue does not yield a preparation with more than about 50% heavy components when the magnesium concentration of the medium is 5 mM [28].

Another critical factor in the isolation of cerebral polyribosomes is the amount of DOC added to the postmitochondrial supernatant. The final concentration should not exceed 1%. Since DOC and other detergents may sensitize polyribosomes to endogenous RNase or other degradative components [29], the amount of DOC used in the preparation of polyribosomes from cerebral tissue is important. Cerebral polyribosomes are much more sensitive to disruption of the mRNA-ribosome complex by RNase attack or suspension in low magnesium buffer (1 mM) than is the analogous preparation from hepatic tissue [7].

The sucrose density gradient profile of a postmitochondrial supernatant fraction from cerebral cortex is shown in Figure 2. The discontinuous gradient used in the preparation of polyribosomes leaves the material seen at the top of the gradient (1-12 ml) at the 0.5-2.0 M sucrose interface. This material consists of soluble protein, single (80 S) ribosomes, and dimers (two ribosomes). Only the heavy polysomes, as well as a small amount of 80 S and dimer, pass through the 2 M sucrose and are pelleted.

In Figure 3 is shown the sucrose density gradient pattern of a preparation of total polyribosomes from rat cerebral cortex. The analogous hepatic preparation from fasted rats is shown in Figure 4.

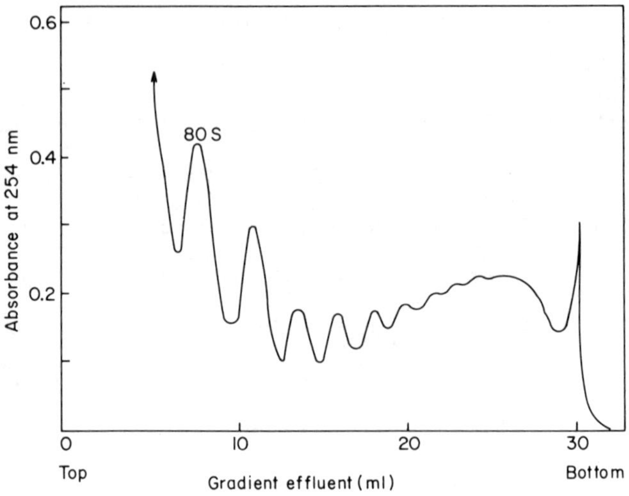

Fig. 2. Sucrose density gradient profile of a postmitochondrial supernatant fraction prepared from cerebral cortex of young adult rats. The postmitochondrial supernatants were prepared in 0.25 M sucrose containing 5 mmoles/liter tris-HCl buffer (pH 7.6), 12 mmoles/liter $MgCl_2$, and 10 mmoles/liter KCl. A 2-ml portion of this fraction was layered on a linear sucrose gradient (15-35%) containing buffer and salts in the concentrations given above. The gradient was centrifuged in a Spinco SW-25 swinging-bucket rotor at 25,000 rpm at 0°C for 4 hr. Absorbance of the effluent collected from the top of the gradient was monitored continuously at 254 nm. The material between the arrows represents the fraction removed during the preparation of the polyribosome fraction.

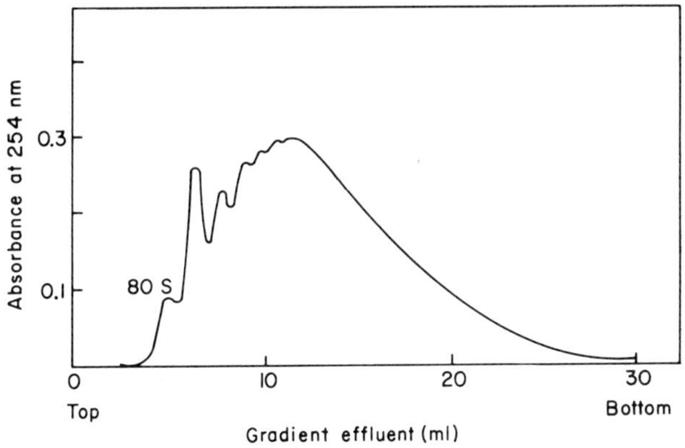

Fig. 3. Sucrose density gradient profile of total polyribosomes prepared from rat cerebral cortex. Polyribosomes were suspended in a medium composed of 0.05 M tris-HCl buffer (pH 7.6) containing 12 mmoles/liter $MgCl_2$ and 0.01 M KCl. A 2-ml portion of this suspension (2.6 A_{260} units) was layered on a linear sucrose gradient (15-35%) containing the same buffer and salts as the suspension medium. The gradient was centrifuged in a Spinco SW-25 rotor at 25,000 rpm at 0°C for 1-1/2 hr. Absorbance of the effluent was monitored continuously at 254 nm.

The method described above can be used for the preparation of total polyribosomes from whole brain, cortex, or hindbrain-medullary region (predominantly white matter). However, cerebral polyribosome preparations from rats always contain a higher proportion of 80 S ribosomes and dimers (10-18%) than isolated hepatic polyribosomes (2-5%). This may be related to

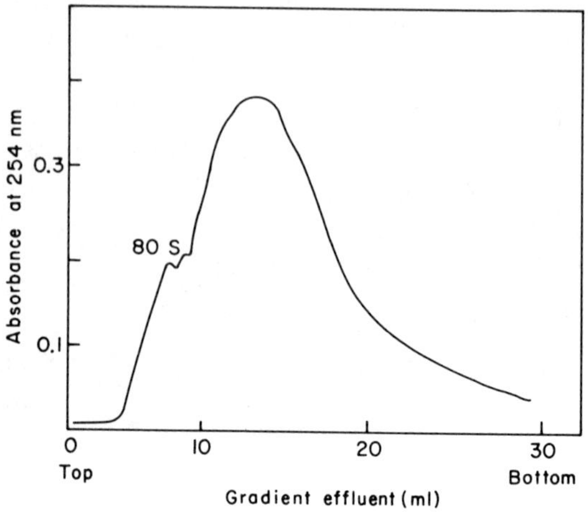

Fig. 4. Sucrose density gradient profile of total hepatic polyribosomes prepared from rats fasted for 18 hr. The procedure for the gradient and subsequent analysis was the same as described in the legend for Figure 3 for cerebral polyribosomes.

---

the apparent instability of the cerebral mRNA-ribosome complex [7], which is manifested at about 14 days of age in the rat cortex [13].

B. Free Polyribosomes

In rat brain cortex, as discussed in Section I, most of the polyribosomes of neuronal cells appear to exist free in the cytoplasm rather than bound to the endoplasmic reticulum. Therefore polyribosomes isolated from rat cerebral cortices without the use of detergents perhaps reflect the situation in vivo and are thus primarily neuronal in origin. The first preparation of free polyribosomes was reported by Campagnoni and Mahler in 1967 [6]. The principle of the method is the same as that for the preparation of total polyribosomes. In both cases the heavier ribosomal components are selectively isolated by centrifuging the postmitochondrial supernatant through 2.0 M sucrose-tris buffer. The procedure presented here is slightly modified from that of Campagnoni and Mahler. The cerebral cortices are homogenized in the following media: 0.25 M sucrose containing 0.05 M tris-HCl (pH 7.6), 0.025 M KCl, and 0.01 M magnesium acetate (or $MgCl_2$). The cortices are minced briefly with scissors in 1 ml of sucrose-tris buffer and homogenized with three to four up-and-down strokes of a Teflon pestle, in a final concentration of 3 ml of medium per gram of cortex. This is sufficient to give a smooth homogenate because of the preliminary mincing procedure. The postmitochondrial supernatant is obtained by centrifugation of the homogenate at 27,000 X g for 10 min. This supernatant fraction is then layered directly over 2.0 M sucrose, containing the same buffer and salts as the homogenizing medium, in a ratio of 1:1. Centrifugation is at 105,000 X g at 0°C for 4 hr. The top layer can be carefully removed to within 2 cm of the 2.0 M sucrose interface and used for isolation of the pH 5 enzymes fraction. This fraction can only be used for preparation of the pH 5 enzymes if free polyribosomes are isolated, since detergent is omitted in this procedure.

Both total and free polyribosomes from rat brain (whether from cortex, white matter, or whole brain) have RNA/RNA plus protein ratios of 0.6. This indicates an RNA content of 60% in the polyribosomes and is similar to the values obtained with preparations from *Escherichia coli*. Most early studies of the RNA/protein ratios from animal cytoplasm reported 40-50% RNA. However, these preparations probably contained nonribosomal protein which remained attached to the ribosomes during the isolation procedure. The use of a higher pH and centrifugation through dense sucrose (2.0 M) removes most if not all of this bound protein.

Polyribosomes that are stable and also active in amino acid incorporation have been prepared from brains of rats ranging in age from newborn (16-18 hr) to adult (42 days).

Sucrose density gradient profiles of postmitochondrial supernatants that have not been treated with detergent, as well as the pattern of the isolated free cerebral polyribosomes, are essentially the same as shown in Figures 2 and 3 for total polyribosomes prepared with detergent.

IV. CELL-FREE AMINO ACID-INCORPORATING SYSTEM

The standard incubation mixture contains the following substances in a final volume of 1 ml: 0.05 M tris-HCl (pH 7.6), 2 mmoles/liter creatine phosphate, 0.1 mg of creatine phosphokinase, a ^{14}C- or ^{3}H-labeled amino acid (or a labeled amino acid mixture), 12 mmoles/liter $MgCl_2$ (or magnesium acetate), 0.10 M KCl, 0.5-1.0 mg of ribosomal protein and 1-4 mg of pH 5 protein [7,14].

NH_4Cl can replace KCl as the source of monovalent cation [6,8]. An ATP-generating system (creatine phosphate-creatine phosphokinase) is obligatory when microsomes or ribosomes are used in the incubation system [14]. However, when amino acid incorporation is carried out with polyribosomes, the dependence on an ATP-generating system appears to be related to the method by which the enzymes used in the incubation system are prepared. If pH 5 enzymes are prepared as described in Section II, C, the generating system is required for a high level of amino acid-incorporating activity <u>in vitro</u>. However, if the pH 5 fraction is prepared from the upper layer of the discontinuous sucrose gradient used in the preparation of free brain polyribosomes, apparently 5 mmoles/liter ATP can replace the requirement for an ATP-generating system [6]. If the 105,000 X g supernatant (cell sap) is used as the enzyme source, rather than the pH 5 fraction, 5 mmoles/liter ATP can be used without the generating system [8,30]. A requirement for the addition of a mixture of 19 amino acids (2 mmoles/ liter with respect to each) minus the radioactive amino acid also seems to be dependent on the method of preparation of the pH 5 or 105,000 X g fractions. If either preparation is dialyzed or passed through a column of Sephadex G-25, free amino acids are removed and a mixture of unlabeled amino acids must be used for maximum incorporation during incubation. Again, if pH 5 enzymes are prepared as described in this chapter, addition of a nonlabeled amino acid mixture actually causes some inhibition when microsomes or ribosomes are used in the cell-free system [14]. With polyribosomes, there is essentially no effect when an amino acid mixture is added to the incubation system since an adequate supply of free amino acids is present in the pH 5 preparation. If the cell sap is used without prior dialysis or passage through Sephadex G-25, there is also no requirement for additional free amino acids.

The amount of ribosomes that must be used in the incubation system depends on the sensitivity of the equipment available for the determination of radioactivity, the specific activity of the isotope used, and the amount of ribosomes and enzyme fractions available to the investigator. The ribosomes we use in the incubation system presented (0.5-1.0 mg) are sufficient to allow sampling of an incubation mixture at several time periods. If only one or two time intervals are needed for an assay, much less material is required. One should always run at least duplicate, and preferably triplicate, samples for any assay. If only a small amount of ribosomes is used (50-200 μg), the final volume of the incubation mixture should not be more than 0.5 ml. However, the ratio of pH 5 enzyme protein to ribosomal protein should be 2:1 for microsomes and ribosomes, and 4:1 for polyribosomes. The higher requirement for the pH 5 enzyme fraction in the polyribosome incubation system is based on the greater purity of this preparation as compared with the isolated microsomes and ribosomes. The latter preparations contain some bound enzymes, thus decreasing the amount of added enzymes required for maximal incorporation.

The creatine phosphate plus creatine phosphokinase system for generating ATP results in a faster rate of endogenous protein synthesis than is provided by the phosphoenolpyruvate kinase system [31].

All the substrates (ATP, GTP, and so on) for cell-free amino acid incorporation are prepared in 0.05 M tris-HCl buffer (pH 7.6) containing 12 mmoles/liter $MgCl_2$ (or magnesium acetate) and 0.10 M KCl. The sample of radioactive amino acid to be used in the incubation system is prepared in the same buffer as the substrates. If the pH 5 enzymes or S-100 fraction have been prepared in buffers containing lower magnesium and/or potassium concentrations, a concentrated solution of each is added so that the final mixture is 0.012 M with respect to Mg^{2+} and 0.10 M with respect to K^+. The particulate preparations (microsomes, ribosomes, or polyribosomes) are suspended in tris-HCl buffer and aliquots are taken for protein determination by the method of Lowry et al. [32]. All components except pH 5 enzymes (or S-100 fraction) and ribosomes are mixed at 0°C. The ribosomes are then added, immediately followed by the enzyme fraction. The latter component is added last since it contains RNase activity [7]. If an artificial messenger such as polyuridylic acid is used, it is prepared in tris-HCl buffer, with the same ion composition as present in the incubation system, and should be added before the ribosomes and enzyme fraction. The final pH of the incubation mixture should be 7.4-7.6 if all the components have been prepared in the same buffer. It is important to keep all materials at 0°C prior to incubation in order to maintain the activity of the cell-free incorporating system. For accurate assessment of the system, a zero-time control should be included in the assay. In this case the reaction is terminated immediately after the addition of all components of the incubation system. Incubation is carried out in air, in a water bath at 37°C with agitation, for the desired time periods. The reaction can be terminated and samples collected, prepared, and counted in several ways (see, for example, Refs. 6,7,30). The procedure used depends on the facilities available to the investigator. The methods described in the references cited above all yield comparable results.

The suspension of microsomes, ribosomes, or polyribosomes from brain, whether for use in cell-free incorporating systems or for analysis by the sucrose density gradient technique, presents some difficulty. Although particulate fractions from most sources tend to slight nonspecific aggregation when suspended in buffer, the extent of aggregation is always greater in preparations from brain. However, a low-speed centrifugation at 5000 X g for 10 minutes is sufficient to clarify the suspension. The loss of material as a result of this nonspecific aggregation is much less for purified polyribosomes than for microsomes or ribosomes, but it still occurs to a greater extent than in comparable preparations from other mammalian sources.

A typical example of the time course of amino acid incorporation by microsomes and ribosomes is shown in Figure 5. The activity of the microsome preparation plateaus after 30 min (Figure 5), or sometimes after 40 min. Similar results have been observed in rat liver [33], as well as brain, by several investigators [14,30,34]. However, it can be seen that the amino acid-incorporating activity of ribosomes continues at a high level for a considerably longer period. This observation has been reported by most investigators [1,14-16,30]. The initial incorporation (the first 30 min of incubation) is actually higher for the microsomal than the ribosomal system if the activity is expressed as counts per minute per milligram of RNA since the microsomes contain only 22% RNA as compared with 45% for the ribosomes. Various interpretations of these results have been presented in the references cited above. At the present time, however, our knowledge of cell-free protein synthesis by preparations from cerebral tissue is not sufficiently

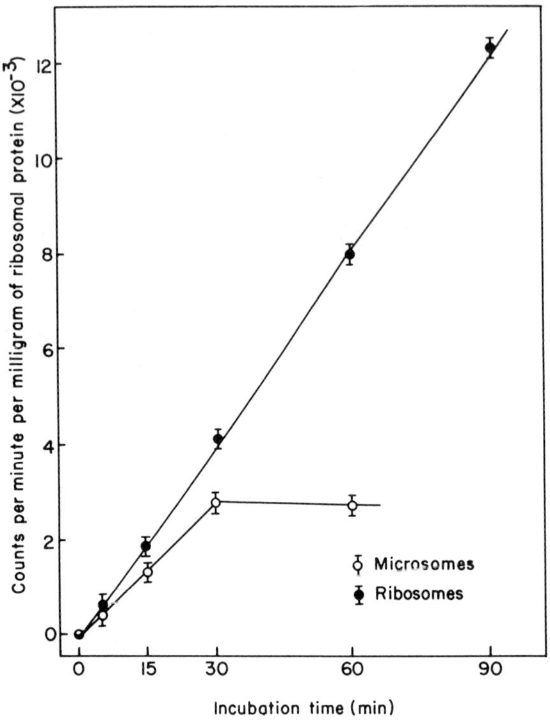

Fig. 5. Kinetics of incorporation of uniformly labeled [^{14}C]L-leucine by cerebral microsomes and ribosomes. The complete incubation system contained 1.0 mg of microsomal or ribosomal protein, 3.0 mg of pH 5 enzyme protein, 2 mmoles/liter sodium creatine phosphate, 0.05 M tris-HCl buffer (pH 7.4), 12 mmoles/liter $MgCl_2$, 0.10 M KCl, and 1 μCi of uniformly labeled [^{14}C]L-leucine in a final volume of 1.0 ml. Incubation took place at 37°C in air. The mean ± S.E.M. is shown for three analyses per time period.

definitive to permit an unqualified statement of the limiting factors.

The amino acid-incorporating activity of polyribosomes prepared from rat cerebral cortex (Figure 6) is always much higher than can be achieved with a ribosome preparation. A direct comparison cannot be made between the extent of incorporation seen in Figures 5 and 6 since the radioactive amino acids differed in type, content, and specific activity. However, if the same concentration (1 μCi) of [^{14}C]phenylalanine with a specific activity of 350 Ci/mole is used in an incubation mixture with cerebral ribosomes, the activity is only one-fifth that obtained when cerebral polyribosomes are used [7]. If Figure 1 (ribosome pattern) is compared with Figure 3 (polyribosomes), it is evident that the latter contains predominantly heavy ribosomal components, whereas the former consists of a high proportion of monomers, dimers, and trimers. This means that a large part of the ribosome fraction cannot take part in the synthesis of any proteins (polypeptides) but can only make di- or tripeptides, since these monomers, dimers, and trimers contain only fragments of mRNA. However, when polyribosomes with a large proportion of functional protein-synthesizing units (aggregates of ribosomes held together by strands of mRNA) are incubated, the completion of proteins occurs at a high level for an *in vitro* system.

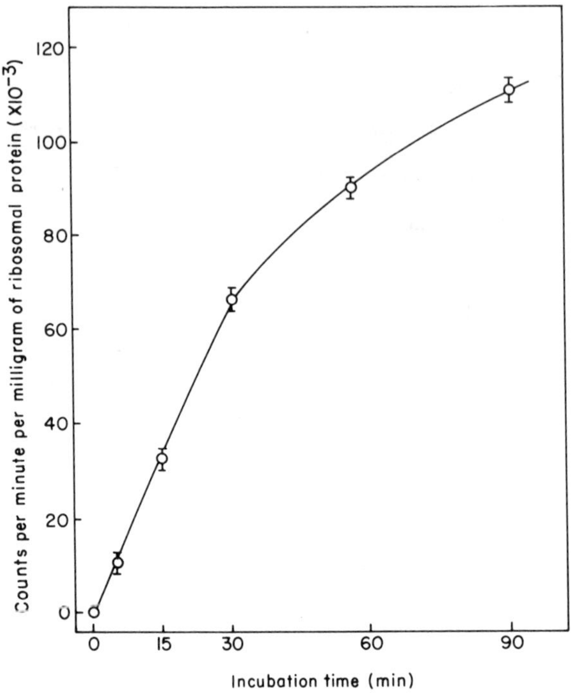

Fig. 6. Kinetics of incorporation of uniformly labeled [^{14}C]L-amino acid mixture by cerebral cortical polyribosomes. The incubation system contained 0.5 mg of polyribosomal protein, 2.0 mg of pH 5 enzyme protein, and 0.25 μCi of [^{14}C]L-amino acid mixture (15 amino acids). The other components and final volume of the incubation mixture are the same as described in the legend for Figure 5 for microsomes and ribosomes. The mean ± S.E. is shown for three analyses at each point.

V. PREPARATION OF RNA WITH PROPERTIES OF mRNA

Some recent theories postulate that memory and learning correlate with quantitative and qualitative changes in the mRNA of the brain [35-37]. However, determination of the content or composition of the total cellular RNA, or even the RNA of subcellular fractions, cannot contribute very much to the elucidation of these proposals since mRNA constitutes less than 5% of the total cellular RNA [38,39]. A method that permits the concentration of mRNA should allow the use of such a mRNA in an *in vitro* system, leading to the synthesis of proteins for which these RNAs were programmed *in vivo*. The observation that the mRNA-ribosome complexes from the cortex of adult rat are relatively unstable has led to development of a method for the isolation of messenger-like RNA species from cerebral cortical polyribosomes [40]. The method employs a discontinuous gradient for the separation of such an RNA fraction after dissociation of cerebral polyribosomes by EDTA.

Polyribosomes, either total or free, prepared by the methods described in the previous sections, are suspended with a glass rod in medium composed of 0.3 mM EDTA (sodium salt), 0.001% polyvinyl sulfate, 0.05 M KCl, and 0.05 M tris-HCl buffer (pH 7.6). About 12 mg of polyribosomal protein are suspended in a final volume of 8 ml of the EDTA-tris buffer. The suspension is allowed to remain for 10 min at 0°C and is then centrifuged for 10 min at 5000 X g at 0°C. The resulting supernatant is layered onto a discontinuous sucrose gradient. The middle layer is composed of a 0.5 M sucrose solution containing the same buffer and other components present in the polyribosomal suspension. The bottom layer consists of 16 ml of 2.0 M sucrose with the same additives. The gradient is centrifuged for 16-17 hr at 0°C at 25,000 rpm in a Spinco SW-25.1 rotor. The whole preparation and gradient can be scaled down to use less material and other swinging-bucket rotors with smaller total capacities. When rotors with greater centrifugal force are used, the time of centrifugation is adjusted accordingly.

After centrifugation the upper layer is carefully removed with a capillary pipet to within 1 cm of a demarcation line that is seen as a cloudy band in the gradient. When larger volumes are used (SW-25.1 tubes), this amounts to about 14 ml of upper layer per tube. This layer is transferred to a 50-ml Erlenmeyer flask, and 3 N NaCl is added to a final concentration of 0.1 N NaCl. After the solution is stirred, 2.5 vol of cold absolute ethanol are added with additional stirring. The mixture is stored at -20°C for 16-18 hr and then centrifuged for 30 min at 27,000 X g at 0°C in Corex tubes with rubber adapters in a Servall SS-34 rotor. This procedure is repeated two to three times; the RNA is concentrated by using smaller volumes (1-2 ml) of EDTA-tris buffer at each successive wash. The final precipitate is dissolved in 0.05 M tris-HCl buffer (pH 7.6) and stored in small aliquots at -60°C at a concentration of 1 mg RNA/ml. The yield is 2% of the total polyribosomal RNA.

Perhaps the most important precautionary measures in the preparation of cerebral mRNA consist of <u>always</u> maintaining all samples at 0°C, centrifuging at this temperature, using RNase-free sucrose, and using double-distilled water. Also, it is advisable to wear disposable vinyl gloves during the procedure to prevent any RNase contamination from the hands.

Direct evidence for an mRNA fraction from brain would be the cell-free synthesis of a brain-specific protein. At the present time, only indirect evidence is available to indicate that this RNA fraction isolated from cerebral polyribosomes may be mRNA. The base ratio [(GMP + CMP)/(AMP + UMP) = 0.87] is close to the value for rat DNA (0.75). DNA-RNA hybridization studies show a large proportion of rapidly labeled molecules within this RNA fraction that hybridize to homologous DNA [40]. Sucrose density gradient analysis of this mRNA fraction (Figure 7) shows two peaks, with sedimentation coefficients approximating 8 and 16 S. The mRNA fraction also exhibits high template activity in a cell-free system containing brain ribosomes (that have been washed with buffer containing a high concentration of salt), pH 5 enzymes, and crude ribosomal initiating factors (Figure 8). The incorporating activity of this natural messenger-like RNA is relatively low compared, for example, with a polyribosome cell-free amino acid-incorporating system. A possible reason for this difference is the presence of RNase activity in the crude initiating factor preparation and the pH 5 enzymes. This would have a more deleterious effect on a cell-free system containing unprotected mRNA as compared with a polyribosome system, in which the mRNA is in a

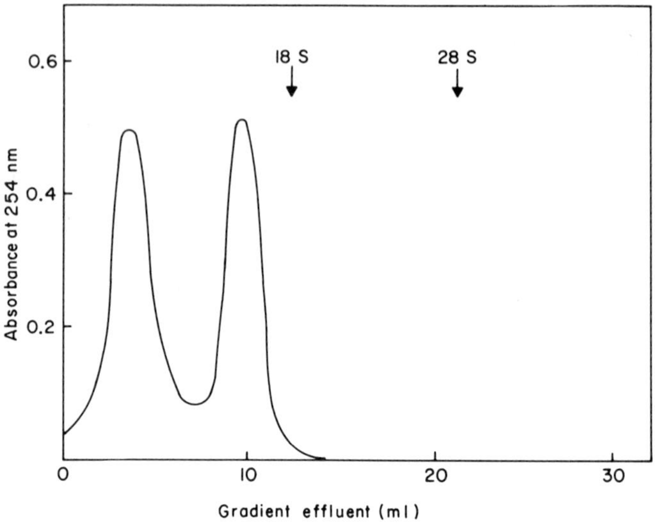

Fig. 7. Sucrose density gradient profile of the polyribosomal mRNA fraction from cerebral cortex of adult rat. The RNA preparation (300 μg) was dissolved in 2 ml of buffer composed of 3 mmoles/liter EDTA, 10 mmoles/liter NaCl, and 0.10 M sodium acetate (pH 5.2). The solution was layered onto a linear sucrose gradient (5-20%) containing the same buffer and salts as the original solution. The gradient was centrifuged in a Spinco SW-25.1 swinging-bucket rotor at 25,000 rpm at 0°C for 18 hr. The effluent from the top of the gradient was monitored continuously for absorbance at 254 nm.

---

relatively protected state as part of the polyribosome complex. Other possible limiting factors are a deficiency of transfer enzymes and chain initiating, as well as chain terminating, factors. The polyribosomes, as isolated, contain polypeptides

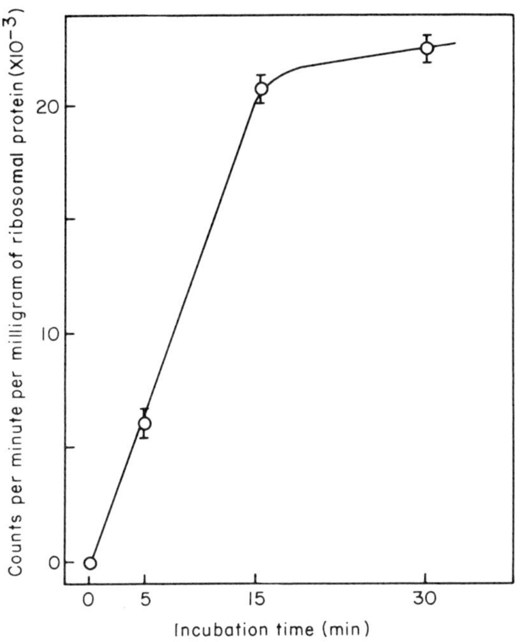

Fig. 8. Kinetics of amino acid incorporation by cerebral ribosomes (high-salt-treated) in the presence of polyribosomal mRNA fraction. The incubation mixture contained 0.5 mg of ribosomal protein, 2 mg of pH 5 enzyme protein, 7 mmoles/liter $MgCl_2$, 0.10 M KCl, 60 μg of crude ribosomal factor protein, 150 μg of the mRNA fraction, 0.25 μCi of a mixture of 15 [^{14}C]L-amino acids, and the other additives described in Section IV. Each value represents the average ± S.E. for three determinations.

---

in various stages of elongation, and much of the protein synthesis observed *in vitro* represents completion of existing chains rather than *de novo* synthesis of proteins [8]. An additional factor to be considered is the possibility that there is a limit to the amount of mRNA that can be bound to the ribosome to form the active unit. The ribosomes used in the *in vitro* system with natural mRNA are converted to 80 S particles by the high-salt treatment. It appears from studies with active subunits (40 and 60 S) obtained from reticulocytes and liver [31,41,42] that mRNA binds to the small subunit (40 S) prior to combination with the large subunit (60 S) to form the 80 S ribosome. Preliminary results (unpublished data) with active subunits from cerebral tissue indicate that a similar situation exists in brain. Therefore the use of purified enzymes and protein factors, as well as purified active subunits from brain, should result in a high level of *de novo* protein synthesis directed by natural cerebral mRNA. These studies are now in progress in this laboratory but are still in a preliminary stage and therefore not appropriate for inclusion in this chapter.

The ribosome preparation used in testing cerebral mRNA *in vitro* is useful, however, for preliminary testing as well as for assaying for initiation steps [44] and for purity of chain initiation and transfer factors. The procedure for preparation of 80 S ribosomes relatively free of endogenous mRNA, as well as bound factors and enzymes, is presented in the next section.

VI. PREPARATION OF "STRIPPED" RIBOSOMES

Cerebral ribosomes, prepared as described in Section II, are rinsed and then suspended in medium containing 0.5 M NH_4Cl, 1 mmole/liter $MgCl_2$, and 0.05 M tris-HCl buffer (pH 7.6) in an amount equivalent to 0.6 g of tissue per milliliter. The suspension is centrifuged for 10 min at 5000 X g at 0°C. The supernatant is removed and allowed to remain for 60 min at 0°C prior to centrifugation for 10 min at 5000 X g at 0°C. The resulting supernatant is then layered on a discontinuous sucrose gradient which contains the same buffer and salts. The final gradient is composed of three layers in the ratio of 1:1:1. The top layer consists of the sample containing the salt-treated ribosomes, a middle layer consists of 0.5 M sucrose, and a bottom layer consists of 1.0 M sucrose. Both sucrose layers contain the same buffer and salts as the ribosome suspension. The gradient is centrifuged for 6 hr at 165,000 X g at 0°C. The resulting pellets are rinsed three times in medium composed of 1 mmole/liter $MgCl_2$ and 50 mmoles/liter tris-HCl buffer (pH 7.6) and are stored at -60°C [40]. A method for the preparation of "stripped" ribosomes from calf brain has been recently reported [45]. The various procedures described, as well as the references cited, contain sufficient information for investigators entering the field of cerebral protein synthesis to use for their particular research requirements. Whether one is interested in the specific mechanisms in brain, or more general changes induced by various treatments of test animals, the available methods presented herein are those that have proved to be consistently reproducible in our laboratory.

REFERENCES

[1] M. R. V. Murthy and D. A. Rappoport, Biochim. Biophys. Acta, 95, 121 (1965).

[2] D. H. Clouet, M. Ratner, and N. Williams, Biochim. Biophys. Acta, 123, 142 (1966).

[3] K. H. Stenzel, R. F. Aronson, and A. L. Rubin, Biochemistry, 5, 930 (1966).

[4] M. K. Campbell, H. R. Mahler, W. J. Moore, and S. Tewari, Biochemistry, 5, 1174 (1966).

[5] C. E. Zomzely, S. Roberts, D. M. Brown, and C. Provost, J. Mol. Biol., 20, 455 (1966).

[6] A. T. Campagnoni and H. R. Mahler, Biochemistry, 6, 956 (1967).

[7] C. E. Zomzely, S. Roberts, C. P. Gruber, and D. M. Brown, J. Biol. Chem., 243, 5396 (1968).

[8] H. R. Mahler and B. J. Brown, Arch. Biochem. Biophys., 125, 387 (1968).

[9] M. P. Lerner and T. C. Johnson, J. Biol. Chem., 245, 1388 (1970).

[10] R. Ekholm and H. Hydén, Ultrastructure Res., 13, 269 (1965).

[11] E. G. Gray, in Electron Microscopic Anatomy (S. M. Kuntz, ed.), Academic Press, New York, 1964, 407.

[12] I. Merits, J. C. Cain, E. J. Rdzok, and F. N. Minard, Experientia, 25, 739 (1969).

[13] C. E. Zomzely, S. Roberts, S. Peache, and D. M. Brown, J. Biol. Chem., 246, 2097 (1971).

[14] C. E. Zomzely, S. Roberts, and D. Rappoport, J. Neurochem., 11, 567 (1964).

[15] G. Acs, A. Neidle, and H. Waelsch, Biochim. Biophys. Acta, 50, 403 (1961).

[16] S. C. Bondy and S. V. Perry, J. Neurochem., 10, 603 (1963).

[17] R. K. Datta and J. J. Ghosh, J. Neurochem., 10, 363 (1963).

[18] C. E. Zomzely, S. Roberts, D. M. Brown, and C. Provost, J. Mol. Biol., 20, 455 (1966).

[19] J. W. Littlefield and E. B. Keller, J. Biol. Chem., 224, 13 (1957).

[20] E. B. Keller and P. Zamecnik, J. Biol. Chem., 221, 45 (1956).

[21] J. Mansbridge and A. Korner, Biochem. J., 89, 15P (1963).

[22] A. J. Munro, R. J. Jackson, and A. Korner, Biochem. J., 92, 289 (1964).

[23] R. W. Risebrough, A. Tissières, and J. D. Watson, Proc. Natl. Acad. Sci. U.S., 48, 430 (1962).

[24] J. R. Warner, A. Rich, and C. E. Hall, Science, 138, 1399 (1962).

[25] A. Gierer, J. Mol. Biol., 6, 148 (1963).

[26] S. Penman, K. Scherrer, Y. Becker, and J. E. Darnell, Proc. Natl. Acad. Sci. U.S., 49, 654 (1963).

[27] F. O. Wettstein, T. Staehelin, and H. Noll, Nature, 197, 430 (1963).

[28] Y. Takahashi, K. Mase, and H. Sugano, Biochim. Biophys. Acta, 119, 627 (1966).

[29] E. L. Kuff and N. E. Roberts, J. Mol. Biol., 26, 211 (1967).

[30] A. J. Dunn, Biochem. J., 116, 135 (1970).

[31] A. K. Falvey and T. Staehelin, J. Mol. Biol., 53, 1 (1970).

[32] O. H. Lowry, N. J. Rosebrough, A. L. Farr, and R. J. Randall, J. Biol. Chem., 193, 265 (1951).

[33] A. Korner, Biochem. J., 81, 168 (1961).

[34] Y. Takahashi, K. Mase, and S. Abi, J. Biochem. (Japan), 60, 363 (1966).

[35] C. W. Dingman and M. B. Sporn, J. Psychiatric Res., 1, 1 (1961).

[36] H. Hydén and E. Egyhazi, Proc. Natl. Acad. Sci. U.S., 49, 618 (1961).

[37] H. Hydén, Progr. Nucleic Acid Res. Mol. Biol., 6, 187 (1967).

[38] R. Soeiro, M. H. Vaughan, J. R. Warner, and J. E. Darnell, Jr., J. Cell Biol., 39, 112 (1968).

[39] Z. S. Tencheva and A. A. Hadjiolov, J. Neurochem., 16, 769 (1969).

[40] C. E. Zomzely, S. Roberts, and S. Peache, Proc. Natl. Acad. Sci. U.S., 67, 644 (1970).

[41] S. Bonanou, R. A. Cox, B. Higginson, and K. Kanagalingam, Biochem. J., 110, 87 (1968).

[42] T. E. Martin and I. G. Wool, J. Mol. Biol., 43, 151 (1968).

[43] A. K. Falvey and T. Staehelin, J. Mol. Biol., 53, 1 (1970).

[44] S. S. Keiwar, C. Spears, and H. Weissbach, Biochem. Biophys. Res. Commun., 41, 183 (1970).

[45] F. Goodwin, D. Shafritz, and H. Weissbach, Arch. Biochem. Biophys., 130, 183 (1969).

Chapter 7

PROTEIN BIOSYNTHESIS IN *Paramecium* WITH SPECIAL REFERENCE TO THE *IN VITRO* SYNTHESIS OF THE CELL SURFACE ANTIGENS

John Sommerville

Department of Zoology
University of St. Andrews
St. Andrews, Fife
Scotland

I. INTRODUCTION 190

II. GENERAL PROCEDURES IN THE PREPARATION OF *Paramecium* CELL FRACTIONS 193

A. Cultures 193

B. Harvesting and Homogenization. 195

C. Preparation of Crude Supernatants. 196

D. Isolation of Ribosomes 198

E. Preparation of pH 5 Fractions. 199

F. Sucrose Gradient Separations 201

III. AMINO ACID INCORPORATION BY *Paramecium* CELL-FREE SYSTEMS. 205

A. Cell-free Incubation and Assay of Protein Synthesis205

B. Effect of Mg^{2+} Concentration 207

C. Incorporation Kinetics 208

D. Effect of Inhibitors 211

IV. DETECTION AND ASSAY OF LABELED SURFACE ANTIGEN PROTEINS 213

A. Preparation of Antigens and Antibodies 213

B. Immunological Assay of Antigen Biosynthesis. . . . 216

V. METHODS OF TRACING THE BIOSYNTHETIC PATHWAY OF SURFACE ANTIGEN PROTEINS 221

A. Synthesis by Isolated Gradient Fractions 221

B. Location of Nascent Antigen by the Use of Labeled Antibodies 223

C. Membrane-Ribosome Associations 224

D. Transport and Cell Surface Appearance of Antigen Protein. 225

VI. OTHER PROTOZOAN SYSTEMS. 226

REFERENCES 227

I. INTRODUCTION

Protein biosynthetic systems derived from protozoan cells have received relatively little attention in spite of the fact that a cell-free amino acid-incorporating system derived from the ciliate *Tetrahymena pyriformis* was described by Magar and Lipmann [1] as early as 1958. In recent years protein synthesis in the related ciliate *Paramecium aurelia* has been studied by research groups at the C.S.I.R.O., Division of Animal Genetics, Epping, New South Wales, Australia [2], and at the Protozoan Genetics Unit, Institute of Animal Genetics, Edinburgh, Scotland [3-6]. Most of the methods described and discussed in this chapter are those that have been routinely used by Dr. R. E. Sinden and myself in the Edinburgh laboratory.

First, the suitability of *Paramecium* in providing a profitable system for the study of protein synthesis should be critically considered. Although *Paramecium* is in one sense a eukaryotic cell, in another sense it is an acellular organism and as such occupies what may be a strategically important level of organization between prokaryotic cells and the cells of multicellular organisms. Does this reflect itself in the conditions and mechanisms of protein biosynthesis? Furthermore, as an unspecialized cell, that is, an acellular organism, *Paramecium* has no specialized function to synthesize substantial amounts of any specific protein. This situation contrasts with that of the more exploited eukaryotic systems, such as those derived from reticulocytes and plasma cells, which are specially adapted to produce considerable quantities of hemoglobin and immunoglobulin, respectively. Characterization of the products of biosynthetic systems is of great importance in establishing synthesis as "genuine" (i.e., the *de novo* formation of recognizable polypeptides) and permits a study in depth of specific biochemical control mechanisms.

My own involvement in the *Paramecium* system arose not merely as a study of protein synthesis *per se*, but through a special interest in the way that protein synthetic techniques, particularly cell-free systems, allow one to tackle the more general problem of cell differentiation, that is, control of the biosynthesis of specific cell proteins. Such an approach would have proved difficult were it not for the presence in *Paramecium* of a category of proteins, the cell surface proteins, that have highly specific antigenic properties and so are amenable to sensitive immunological assay techniques.

We hope, therefore, that this chapter serves a 3-fold purpose: (1) to describe the conditions for cell-free protein synthesis in *Paramecium* and to indicate how protozoa may resemble or differ from prokaryotic and other eukaryotic systems; (2) to describe techniques used in the immunological assay of *de novo* synthesis of small amounts of antigenic protein; this, we hope, may be of general interest in the study of those antigenic proteins that constitute a minor fraction in other organisms; and (3) to illustrate how labeling methods used in conjunction with immunological techniques can be used to follow the biosynthetic process from the initial site on the ribosomes to the final site on the cell surface. There is no doubt that biosynthetic mechanisms and intracellular transport are intimately connected processes, and we are only now beginning to understand how these processes are integrated and controlled. The nature of the control of expression of cell surface antigens is a key problem in many fields of biology today.

II. GENERAL PROCEDURES IN THE PREPARATION OF *Paramecium* CELL FRACTIONS

A. Cultures

Paramecium aurelia can be cultured in a number of different media including nonaxenic bacterial broths, semidefined axenic media, and completely defined media. (For full details refer to Sonneborn [7] and Soldo and Van Wagtendonk [8].) These various culture methods each have their own particular advantages and disadvantages. Feeding paramecia on bacteria, their natural diet, facilitates good growth rates (from one fission per day at 18°C to four or five fissions per day at 31°C). The obvious drawback to this procedure is the possible contamination of paramecium preparations with bacteria and bacterial cell components. This factor becomes especially significant in relation to the examination of subcellular fractions and the performance of biochemical analyses. However, this problem has proved to be not a particularly serious one if the following precautions are taken: (1) the paramecium cells are washed as free as possible of bacteria by using slow-speed centrifugation; (2) homogenization is relatively gentle, preventing disruption of bacteria; and (3) postmitochondrial (i.e., postbacterial) supernatants are used as a starting point in the preparation of components to be utilized in biosynthesis experiments. These precautions are elaborated upon later.

Although axenic culture methods overcome the above problems, they too have their disadvantages in that growth is generally poor (never exceeding one fission per day). In addition, only certain stocks can as yet be maintained with any success on axenic media.

For batch cultures the method we find most successful is that described by Jones [9]. Dried grass (obtainable from J. McIntyre, Albert Mills, Easter Road, Edinburgh) is added to distilled water at a concentration of 100 g/liter, and an infusion is prepared by boiling the mixture for 15 min. The extract is filtered through muslin and sterilized by autoclaving at 20 lb/in^2 for 15 min. As required, working-strength medium is prepared by diluting the grass extract 1:20 with distilled water and adjusting to pH 6.8 with Na_2HPO_4. This medium is then dispensed in 1-liter portions in Thompson bottles and autoclaved at 20 lb/in^2 for 15 min. After inoculating the medium with *Klebsiella aerogenes*, growth is initiated by adding at least 50 ml of stationary-phase *P. aurelia*. The bottles are then stacked flat to give a culture depth of no more that 3-4 cm. For G serotype organisms of syngen 1 stocks, the culture temperature is maintained at 18 or 24°C, and for D serotypes at 31°C. An inherent danger in this type of culture is an overproliferation of the bacterial population, which tends to have a detrimental effect on the paramecia. Therefore a balance must be established whereby the bacteria are not allowed to outgrow the paramecia. This is normally accomplished by adding only a small inoculum of bacteria when starting a new culture.

B. Harvesting and Homogenization

Paramecia are ideally harvested at late log phase when the medium has nearly cleared of bacteria and paramecium fission stages are still present. After they are filtered through a layer of absorbant cotton wool over muslin to remove debris, the cells are concentrated by centrifuging in 125-ml, pear-shaped bottles at 500 X g for 2 min in an oil-testing centrifuge. The supernatant fluid is aspirated off, and the cell pellets are pooled, washed in isotonic buffered salts solution [13 mM NaCl, 3 mM KCl, 3 mM Ca_2Cl, 4 mM phosphate buffer (pH 6.8)], and recentrifuged. Washing and centrifugation is repeated until a pure-white cell pellet is obtained and the supernatant shows no traces of bacterial contamination. The cell pellet is finally suspended in diluted homogenization buffer [0.25 M sucrose, 0.1 M tris-HCl (pH 7.6), 0.05 M KCl, 10 mM $MgCl_2$, 10 mM 2-mercaptoethanol, diluted 10-fold with ice-cold distilled water], and centrifuged at 1000 X g for 2 min. The supernatant is completely drained off by the use of a Pasteur pipet attached to a water suction pump. Generally,

2-3 liters of culture yield 1 ml of packed cells, which corresponds to approximately 10^7 organisms.

To 1 vol of packed washed cells, 3 vol of ice-cold homogenization buffer are added. All subsequent procedures should be carried out at 0-4°C. Although the cell cortex of *Paramecium* is relatively tough, the large size of the cells (approximately 150 X 50 μm) makes lysis by shearing force relatively easy. A suitable mechanical homogenizer is a Tri-R Teflon homogenizer which is operated at speed 6 for about 20 strokes or until over 95% of the cells can be seen to be lysed on microscopic examination. Manual homogenization can also be employed, but repeatable conditions are less easy to determine. Detergent treatment of harvested cells should be avoided because the lysate is subsequently unable to produce an active amino acid-incorporating system. Detergents used have been 0.5% solutions of sodium deoxycholate (DOC), Tween 80, Brij 58, and Nonidet P-40. Nonionic detergents do not cause visible damage to cell organelles, such as nuclei and mitochondria, and presumably the failure to produce an active protein-synthesizing cell-free system is attributable to the solubilization of degradative enzymes.

C. Preparation of Crude Supernatants

The homogenate is partially clarified by centrifuging at 2000 X g for 10 min. The supernatant is carefully pipetted out of the centrifuge tubes, avoiding disturbance of the pelleted material and the floating lipid layer. The supernatant is centrifuged this time at 9000 X g for 10 min, and the same precautions are taken in removing the middle supernatant fraction. This supernatant is still noticeably turbid but should contain no visible particles, such as bacteria or mitochondria, on examination of sample with a phase-contrast microscope. This fraction is referred to as the postmitochondrial fraction or S-9. The turbidity can be removed by centrifuging the postmitochondrial supernatant at 20,000 X g for 20 min and pipetting off the top two-thirds of the supernatant. This supernatant has been shown by electron microscopy to be composed solely of ribosomes and soluble factors and is referred to as S-20. The pellet has been shown to consist of various membranous vesicles, including structures analogous to rough and smooth endoplasmic reticulum, as well as considerable quantities of glycogen particles. This membrane fraction, or P-20, has been used in some incorporation studies and can be prepared by carefully rinsing the pellet, resuspending it in homogenization buffer, and recentrifuging at 20,000 X g for 20 min. This time the supernatant wash is completely discarded and the pellet is resuspended in dialysis buffer [50 mM tris-HCl (pH 7.6), 25 mM KCl, 5 mM $MgCl_2$, 5 mM 2-mercaptoethanol]. Prior to amino acid incorporation studies, all crude fractions are dialyzed for 2-4 hr at 0°C against two changes of at least 100 vol of dialysis buffer.

D. Isolation of Ribosomes

Ribosomes can be prepared from postmembrane supernatants (S-20) by centrifugation at 110,000 X g for 90 min. The upper two-thirds of the supernatant is pipetted off and dialyzed against dialysis buffer for 2-4 hr at 0°C before use (or stored at -20°C where it remains active for several weeks). This is referred to as soluble fraction, or S-110. The remaining supernatant is discarded, and the clear pellet is surface-rinsed several times with ice-cold dialysis buffer and drained by inverting the centrifuge tube. The walls of the tube can be carefully wiped dry with tissue paper and cut away to leave only the bottom portion of the tube containing the now-accessible, clear, ribosomal pellet, P-110. This pellet is scraped from the tube with a microspatula, and the material is resuspended by gentle manual homogenization in a minimal volume of dialysis buffer. Precipitated or nonresuspended material is removed by centrifugation at 20,000 X g for 20 min. The supernatant is pipetted off, adjusted to a final concentration of about 50 A_{260} units/ml, and kept at 0°C if it is to be used within a few hours. Otherwise, the ribosomal preparation is stored at -20°C. All these operations should be carried out quickly in a cold room at 0-4°C, with all containers in ice.

The cell fractionation operations described above are summarized in the accompanying flow diagram (Figure 1).

E. Preparation of pH 5 Fractions

Unsuccessful attempts have been made in different laboratories to obtain a paramecium pH 5 fraction which would stimulate amino acid incorporation by ribosomes. It was found that pH 5 fractions derived from *Paramecium* caused a significant inhibition of amino acid incorporation by mouse liver microsomes [2] and rabbit reticulocyte ribosomes [3]. However, pH 5 fractions prepared from mammalian cells have been found to stimulate incorporation by *Paramecium* ribosomes. Furthermore,

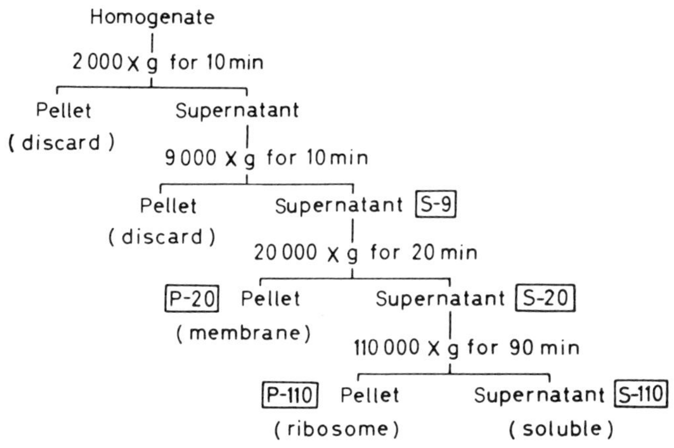

Fig. 1. Flow diagram showing derivation of cell fractions from a paramecium homogenate.

ribosome-free supernatants derived from mouse liver cells have been reported to give a 4-fold increase in the synthetic capacity of Paramecium ribosomes over a system containing an equivalent paramecium supernatant [2] (see Figure 3C).

The pH 5 fraction from mouse liver is prepared by a method similar to that described by Reisner and Macindoe [2]. Freshly excised livers are transferred to a petri dish containing ice-cold buffer [0.25 M sucrose, 50 mM tris-HCl (pH 7.6), 25 mM KCl, 5 mM $MgCl_2$, 5 mM 2-mercaptoethanol] at a concentration of 2-3 ml buffer per gram wet weight of liver. The livers are finely diced with a razor blade, transferred to a glass tube, and homogenized manually with a Teflon pestle until the tissue is totally disintegrated. After preliminary low-speed centrifugation to remove the large debris, the clarified homogenate is centrifuged at 140,000 X g for 60 min. The upper two-thirds of the supernatant is diluted 4-fold with buffer; then the pH of the solution is adjusted to 5.2 by the dropwise addition of ice-cold 1 N acetic acid with continuous stirring. The precipitate formed is collected by centrifugation at 10,000 X g for 10 min, and the pellet is surface rinsed with the pH 7.6 buffer minus sucrose and resuspended in this solution to give a final concentration of 8-10 mg/ml. The pH 5 fraction is kept on ice until ready for use, or stored at -20°C where it remains active for a few weeks.

F. Sucrose Gradient Separations

Postmitochondrial supernatants, ribosome suspensions, or membrane fractions can be further separated into more discrete and purified subcellular components such as rough and smooth membranes, polyribosomes of various sizes, monosomes, and soluble material. This is achieved mainly be means of sucrose density gradient centrifugation, although agarose gel filtration (Sepharose 4B, Pharmacia) has recently been used to separate monosomes and polyribosomes [10]. The separated components are analogous to various structures seen on examining the cytoplasmic ultrastructure of *Paramecium* in sections of whole cells. *In situ*, the majority of the ribosomes appear to be free in the cytoplasm, either singly, or arranged in linear or helical aggregates of various sizes, while the remainder appear to be attached to membranous structures resembling the rough endoplasmic reticulum of mammalian cells. Smooth membranous structures are also apparent. In purifying the various subcellular fractions, the design of the gradient is important (see Figure 2).

For crude supernatants, or any preparation containing a membrane fraction, gradients are prepared by first introducing into the bottom of a 35-ml centrifuge tube a 6-ml cushion of 60% (w/v) sucrose to trap rapidly sedimenting material and avoid pelleting. This is overlayed with a 10-25% (w/v) linear sucrose gradient of 25-ml total volume. These volumes are suitable for an M.S.E. 40-ml swing-out (swinging-bucket) rotor and are adjusted proportionately for other rotor heads. All sucrose solutions are made up to contain a final concentration of 50 mM tris-HCl (pH 7.6), 25 mM KCl, 5 mM $MgCl_2$, and 5 mM 2-mercaptoethanol, and are kept at 0-4°C. The gradients are made using a two-chambered gradient maker with the mixing and delivery chamber containing the denser solution. A fine stream of decreasing density sucrose is delivered through 0.6-mm-bore Teflon tubing and allowed to run down the inner wall of the centrifuge tube. As the gradient is formed, the end of the delivery tube is moved up gradually to remain always within several millimeters of the surface. The time required to form a 25-ml gradient should be about 30 min. This procedure is best performed in a cold room operating at under 4°C where the finished gradient tubes are allowed to remain for 2-4 hr before use in order for the gradients to equilibrate completely.

One-milliliter samples (each containing 25-30 mg of protein in the case of a dialyzed S-9 preparation) are carefully layered on top of the gradients. Centrifugation is for 2-2.5 hr at 20,000 rpm (g_{av} = 45,000) at 2°C. The bottoms of the tubes are punctured with a gradient piercer, and the contents are pumped out using a peristaltic pump at a flow rate of 0.5-1 ml/min. The effluent flow is passed through an ultraviolet spectrophotometer scanner (LKB Uvicord II), and the absorbance profile at 254 nm is continuously recorded on a potentiometric chart recorder (Figure 2A). Finally, 1- or 2-ml fractions are collected with a drop-counting fraction collector (LKB 7000 UltroRac Fractionator). The elution of the gradients is performed in a cold room.

For purified ribosomes (P-110), the preparation and analysis of the gradients is similar to that described above

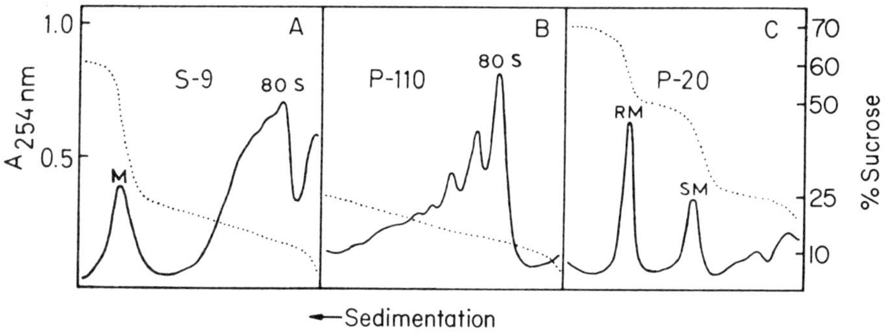

Fig. 2. Methods by which different sucrose gradients can be used to separate ribosomal fractions for various types of incorporation study: (A) 60% sucrose cushion, 10—25% sucrose gradient, for analysis of membrane fraction (M) and ribosomes of S-9. (B) 10—25% sucrose gradient for separation of discrete ribosomal peaks. (C) 70—50—25% sucrose step gradient for separation of rough membrane (RM) from smooth membrane (SM).

except that the dense sucrose cushion is unnecessary since membranous material, if present as a contaminant, is best pelleted on the bottom of the tube. The total linear gradient volume is now increased to 31 ml (or to a few milliliters less than the capacity of the tube). The sample applied should contain between 5 and 20 A_{260} units/ml, and centrifugation time can be lengthened to 4 hr (at 45,000 X g) to give a better separation of discrete polyribosome peaks (see Figure 2B).

For further separation of the membrane fraction (P-20), a useful procedure is to make a step gradient of 6 ml of 70% sucrose (w/v), 6 ml of 50% sucrose (w/v), and 8 ml of 25% sucrose (w/v) (or in these respective proportions by volume). On top of this is layered the resuspended membrane fraction, and the tubes are centrifuged for 2 hr at 30,000 rpm (100,000 X g_{av}) in an M.S.E. 3 X 20 ml rotor. Two major peaks are formed: one at the 70-50% interface and a second at the 50-25% interface (see Figure 2C). Electron microscope examination of the material contained in these peak fractions shows that the denser material consist of predominantly membrane-bound ribosomes, while the less dense material contains smooth membranous vesicles. Other minor fractions are found in the upper regions of the gradient and consist of slower-sedimenting ribosomes and floating lipid. The purified membrane-bound ribosome fraction can be used in amino acid incorporation studies.

III. AMINO ACID INCORPORATION BY *Paramecium* CELL-FREE SYSTEMS

A. Cell-Free Incubation and Assay of Protein Synthesis

The various ribosome-containing preparations described above can be tested for their capacity to incorporate amino acids into protein. The conditions for optimal incorporation are detailed below. It can be seen that cell-free systems derived from *Paramecium* are broadly similar to those described for other organisms in requiring GTP, Mg^{2+}, and an energy-generating source for protein synthesis.

The incorporation mix contains in a total of 1 ml: 50 μmoles tris-HCl (pH 7.6), 25 μmoles KCl, 5 μmoles $MgCl_2$, 5 μmoles 2-mercaptoethanol, 0.2 μmole GTP, 0.2 μmole ATP, 5 μmoles creatine phosphate, 20 μg of creatine phosphokinase, and 10 μmoles each of 18 amino acids. Radioactive labeling is provided for by substituting for the corresponding nonradioactive analogs 1 μCi of either [^{14}C]amino acids or [^{35}S]amino acid in the form of [^{14}C]L-leucine (165 Ci/mole), [^{35}S]L-cysteine (>100 Ci/mole), or an [^{14}C]amino acid mixture (45 Ci/atom of carbon), all obtained from the Radiochemical Center, Amersham. The biosynthetic factors are added as 0.2-0.4 ml of dialyzed crude supernatant, S-9 or S-20 (containing about 3 mg of protein), or as an equivalent amount of a mixture of purified ribosomes and S-110 or pH 5 fraction. Samples from gradient fractions vary in their ribosomal content, and to each of these is added an equal amount of soluble material from the top gradient fraction. Normally, 0.1 ml of soluble fraction is added to each 1.5-ml aliquot of the other gradient fractions, and an incorporation mix is made up to a total volume of 2 ml.

Exceptions to this treatment are the membrane fractions that require an excess of soluble material for optimal activity. The particular ratio of membrane fraction to supernatant is highly critical [5] and should be established for each preparation. Generally, 2 A_{260} units of membrane fraction require 1 ml of gradient supernatant for maximum activity, that is, a mixture approaching that found in the postmitochondrial supernatant. The nature of the components responsible for the inhibitory action in the membrane fractions has not been established, but inhibition could be attributable to endogenous RNase, ATPase, or inhibitors of the amino acid transfer reaction.

All tubes are kept in ice until the incubation is initiated by placing the tubes in water baths at 18, 27, 30, or 35°C. Control tubes are identical except that they contain, in addition, 2 μg/ml of RNase. After the required incubation period, the tubes are plunged into ice, and a 100-fold excess of the appropriate nonradioactive amino acids is added. If the labeled protein is not required for further analysis, the incubates are treated for scintillation counting.

If the protein concentration of the radioactive sample (0.2 ml) is low, 0.1 ml of 1% serum albumin is added as a carrier. To the chilled sample 0.1 ml of NaOH is added to a final concentration of 2 N, and nucleic acids are hydrolyzed by heating the mixture for 2 min at 37°C. The sample tubes are chilled, and 2 ml of 10% (w/v) trichloroacetic acid (TCA) are added to give an excess of acid. After 30 min at room temperature, the precipitates are collected on 2-cm (diameter) membrane filters (Oxoid), washed eight times with 2 ml of 5% TCA, twice with 2 ml of ethanol-ether (1:1), and once with 2 ml of ether. The filters are then dried in a stream of hot air and placed in 20-ml counting vials, each containing 10 ml of scintillation fluid: 0.5% (w/v) PPO and 0.03% (w/v) dimethyl-POPOP in toluene. Radioactivity is measured in a liquid scintillation counter.

B. Effect of Mg^{2+} Concentration

The dependence of amino acid incorporation by *Paramecium* cell-free systems on the concentration of Mg^{2+} is shown in Figure 3A. The concentration of Mg^{2+} is critical, and maximum incorporation occurs at 5 mM. Incorporation is totally suppressed at concentrations of 1 mM or less. (A similar narrow optimum for effective Mg^{2+} concentration has also been reported for a cell-free system derived from the flagellated protozoan *Crithidia oncopelti* [11].) The properties of *Paramecium* ribosomes have been extensively studied by Reisner et al. [12], who have shown that the reduction of amino acid incorporation within the range of 4-1 mM Mg^{2+} is a result of

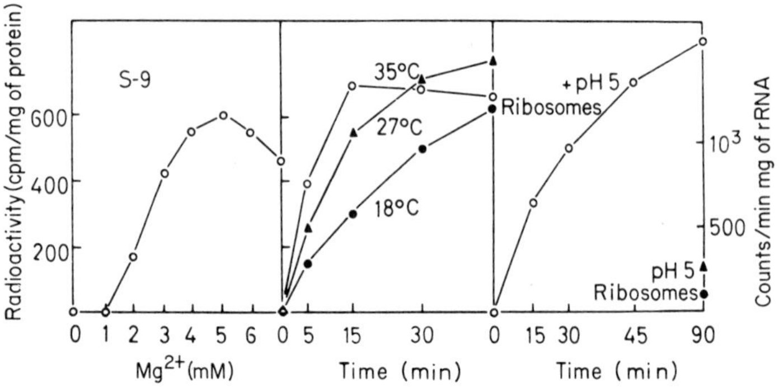

Fig. 3. Incorporation of [^{14}C]amino acids by a paramecium cell-free system: (A) The critical relationship of Mg^{2+} concentration to incorporation. (B) Incorporation kinetics by postmitochondrial supernatants (S-9) at 18, 27, and 35°C. (C) Incorporation kinetics of a mixed system of paramecium ribosomes and mouse liver pH 5 fraction incubated at 27°C.

---

ribosomal disorganization. Ultracentrifuge analysis has shown that over this range there is a progressive change from the 80 S ribosomal form to a more hydrated 60 S form. At 1 mM most of the ribosomes are of the hydrated 60 S form (although no 45 or 30 S subunits are detectable and polyribosomes are still present). These characteristics distinguish *Paramecium* ribosomal systems from other systems and emphasize the importance of maintaining a 5 mM Mg^{2+} concentration at all times. Differences between *Paramecium* and both prokaryotes and eukaryotes are also found in the sedimentation coefficients of ribosomal subunits (25 and 18 S) and in the molecular weights of their RNA (1.25×10^6 and 0.69×10^6) as estimated by rate sedimentation [12] and acrylamide gel electrophoresis [13]. The ciliates appear to most resemble plants in their ribosomal organization.

C. Incorporation Kinetics

Paramecia are cultured under quite a wide range of temperatures, normally between 18 and 35°C. In the cell-free incorporation system there appears to be no real temperature optimum within this range. The only effect of increasing temperature is to speed up the kinetics of incorporation and accelerate the decay in activity. Figure 3B shows the effect of incubation temperature on incorporation by a postmitochondrial supernatant (S-9) derived from cells cultured at 18°C. The final incorporation under the conditions described is maximal at about 700 cpm/mg of protein. Postmembrane supernatants (S-20) give slightly less incorporation (500 cpm/mg of protein) indicating either that the membrane-bound ribosomes are more effective in stimulating amino acid incorporation or that the membrane fraction enhances the total incorporation by interacting with the free ribosomes.

That the membrane fraction is itself active in protein synthesis, by virtue of attached ribosomes, can best be demonstrated by extracting labeled P-20 and treating it with 0.5% sodium DOC; the membrane peak (of A_{254}) on sucrose gradients is thus converted to polysomes, monosomes, and soluble material. Of the radioactive protein over 40% remains in association with the ribosomes, while about 50% is solubilized (see Figure 4C and D). Membrane-released ribosomes are difficult to label *in vitro* because of the release of non-dialyzable inhibitors on detergent treatment.

The mixed system of *Paramecium* ribosomes and mouse liver pH 5 fraction appears to be the most effective combination of factors in incorporating amino acids (Figure 3C). Each 1 ml of incubation mix contains 9 A_{260} units of ribosomes and 0.2 mg of pH 5 protein. The success of this heterologous system indicates that protein synthesis follows the conventional pattern for other eukaryotic systems. However, the nature of the interaction is unknown, since no recognizable newly labeled protein has as yet been isolated from the hybrid system.

Since activity is as prolonged as in many other systems, and the extent of incorporation is quite high (120 nmoles of leucine per milligram), it is strange that there is little release of soluble labeled protein (it amounts to less than 20% of the total labeled protein). After incubation the majority of radioactivity remains associated with the ribosomes, which are now almost entirely in the 80 S form, and with the membrane fraction (see Figure 4B). The breakdown of the system is presumably attributable to endogenous RNase activity. However, various RNase inhibitors such as bentonite, heparin,

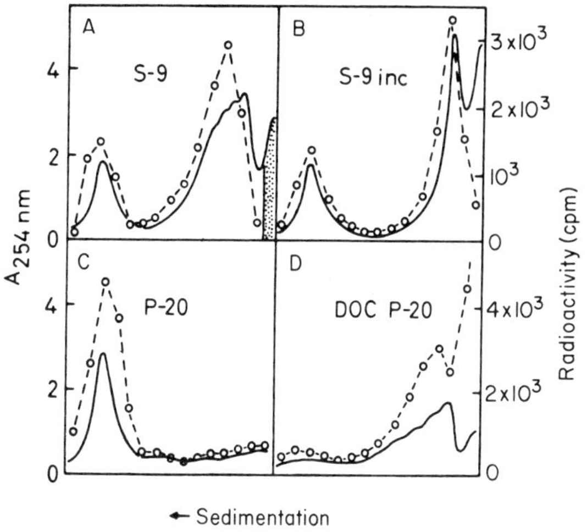

Fig. 4. Sucrose gradient analysis in studying incorporation of [^{14}C]amino acids. (A) Incorporation by isolated fractions derived from an S-9 preparation. The top fraction (stippled) is added to the ribosome-containing fractions as a source of soluble factors. (B) Sedimentation of an S-9 incorporation mixture after incubation at 30°C for 30 min. (C) Sedimentation of isolated membrane fraction (P-20). (D) A_{254} profile and distribution of radioactivity of P-20 after treatment with 0.5% sodium DOC.

---

and polyvinyl sulfate have little or no effect in prolonging activity.

D. Effect of Inhibitors

That the incorporation measured in the Paramecium system is attributable to genuine polypeptide synthesis is checked by observing the effect of known inhibitors of protein biosynthesis.

The addition of low concentrations of RNase (1-2 μg/ml) totally inhibits incorporation by the system and results in the rapid conversion of polyribosomes to 80 S monosomes. This rules out incorporation by contaminating bacteria and is consistent with the hypothesis that a susceptible mRNA binds together a number of monosomes.

The behavior of the cell-free system in the presence of antibiotics should be that expected of an 80 S ribosomal system utilizing preformed mRNA. Actinomycin D, at a concentration of 50 μg/ml, inhibits incorporation only slightly (5%). Chloramphenicol, at concentrations of up to 200 μg/ml, has little effect on incorporation, again demonstrating that bacterial components are not present. The ineffectiveness of chloramphenicol also shows that the ribosomes of Paramecium resemble the ribosomes of other nonbacterial organisms in this respect. Puromycin acts as expected by inhibiting incorporation 63% at a concentration of 200 μg/ml; it can be used to release nascent polypeptide chains from the ribosomes but only while the system is still actively incorporating amino acids. After the incorporation plateaus, puromycin has no releasing powers, and the nascent chains can only be freed by disrupting the ribosomes with 0.02 M EDTA.

However, the only real criterion for genuine synthesis by a cell-free system is a demonstration of the formation of recognizable proteins.

IV. DETECTION AND ASSAY OF LABELED SURFACE ANTIGEN PROTEINS

For several years we have studied the formation of cell surface proteins, the immobilization antigens, which are always present in *Paramecium* and which coat the entire surface area of the cell. The antigen system has many intrinsically interesting properties. The formation of these antigens is subject to a precise genetic control mechanism which permits the expression of only 1 of a range of sometimes more than 12 possible antigens. Which antigen locus is expressed is determined by the particular cytoplasmic factors formed in response to prevailing environmental conditions. In order to investigate the biosynthesis of the surface antigens and the activity of antigen-determining genes, the synthesis of these proteins in the cell-free system has been studied.

A. Preparation of Antigens and Antibodies

Two preliminary procedures must be carried out before immunological assay of protein biosynthesis. The first step is to isolate and purify the required protein; the second is to prepare antibodies specific to this protein.

Concentrated, living paramecia of a single serotype are treated with 4 vol of a solution containing 15% ethanol and 0.045% NaCl for 1 hr at 2-3°C [14]. This treatment breaks off the cilia and solubilizes the surface antigen, yet appears not to lyse the cells. However, a certain amount of leaching of cytoplasmic material into the extract no doubt occurs. The mixture is then centrifuged at 20,000 X g for 10 min to sediment the cells and cilia. This clarified supernatant is then adjusted to 75% saturation with respect to $(NH_4)_2SO_4$ by adding the dry salt, and the solution is stirred overnight at 2-3°C. The precipitate is collected by centrifugation at 10,000 X g for 15 min. It is then resuspended in a minimal volume of distilled water, dialyzed for 48 hr against several changes of distilled water, centrifuged, and finally lyophilized. This crude protein extract is further separated by column chromatography [9]. SE-Sephadex C-50 (Pharmacia), after preliminary washing and swelling in distilled water, is suspended in 0.05 M sodium acetate buffer (pH 4.2). After several washes in this buffer, the slurry is poured into a column to give a packed bed of 1.5 X 15 cm. The resin is then equilibrated by running the starting buffer through the column overnight. The protein is applied as a 2-3% solution, and elution is performed with 800 ml of 0.05 M sodium acetate buffer in a linear gradient of pH 4.2-5.2. The eluted material is monitored by measuring its absorbance at 280 nm with an LKB UviCord II flow spectrophotometer coupled to a chart potentiometer, and fractions are collected. There is always a sharp peak of material that elutes between pH 4.6 and 4.8. The fractions corresponding to this peak are assayed for antigen activity [15]. The fractions containing the bulk of antigen activity are then pooled, dialyzed against distilled water, and lyophilized. In general, 1 ml of packed cells gives a final yield of 0.5 mg of isolated surface antigen. The efficiency of extraction is 60%. The purity of the preparation (as a single protein) can be tested by starch gel [16] or acrylamide gel [17] electrophoresis, electrofocusing [18], or rate sedimentation [9]. Its immunological identity can be checked by double diffusion [19] or immunoelectrophoresis [20].

For the stimulation of specific antibodies, the best method was found to be the multiple emulsion technique of Herbert [21]. This method has the advantages of producing a high and prolonged antibody response but, in addition, unlike the Freund adjuvant, the low viscosity of the inoculate makes injection easy and its water-in-oil-water nature results in dispersed antigen deposits which do not produce localized lesions in the injected animals. This type of emulsion is highly stable and can even be stored at room temperature for many months without evidence of breakdown.

A solution of purified protein antigen dissolved in 0.9% saline is incorporated into a water-in-oil emulsion. Nine parts of Draked 6VR (Pennsylvania Refining Company) are mixed with one part of the emulsifier Arlacle A (Atlas Powder Company) as the oil phase. Equal volumes of oil and protein solution are thoroughly mixed by cycling the mixture through a hypodermic syringe, fitted with a 0.6-mm-bore needle, until a thick viscous emulsion is formed. The water-in-oil emulsion is then reemulsified in an equal volume of 2% Tween 80 in 0.9% saline, and recycling of the mixture through the syringe is continued until a free-flowing liquid is obtained. Production of the required type of multiple emulsion should be confirmed by microscopic examination.

Rabbits 9—12 months of age are injected subcutaneously with between 2 and 5 mg of purified protein antigen contained in a total volume of 2 ml of multiple emulsion. Six weeks after the injection, a blood sample is taken from the marginal ear vein. The blood is allowed to clot, and the serum is drawn off, heated at 56°C for 30 min to inactivate the complement and dialyzed overnight at 2—3°C against 0.9% saline. If a reasonable titer is obtained (activity at over 1:800 dilution [15]), bleedings are continued weekly.

B. Immunological Assay of Antigen Biosynthesis

For the purpose of demonstrating labeled antigenic protein by specific precipitation, an adequate control is necessary to ensure a reliable estimate of nonspecific precipitation. In the *Paramecium* system a homologous antiserum and a heterologous antiserum are used. The homologous antiserum is that prepared against cells of the type used in the experiment; heterologous antiserum is that prepared against the same stock of cells expressing an alternative surface antigen. (We normally work with the 90G-90D system of syngen 1 *P. aurelia*). When such a built-in control cannot be obtained, nonimmune serum can be used as the control.

The most satisfactory serological reactions are obtained by preparing specific antibodies [22]. Globulins are first prepared by repeated precipitation from antiserum with a 50% saturation of $(NH_4)_2SO_4$, then fractionated by either sucrose density centrifugation or ion-exchange chromatography on DEAE-cellulose. The specific antibodies against *Paramecium* surface antigens are of the IgG class, which has a sedimentation coefficient of 7 S, elutes from DEAE-cellulose with 0.01 M phosphate buffer (pH 7.5), and is active in the presence of 50 mM 2-mercaptoethanol [22].

Antigen synthesized in the cell-free system can be assayed by one of two techniques: (1) scintillation counting of specific serological precipitates (this is quantitative but tends to be susceptible to nonspecific effects or (2) autoradiography of immunoelectrophoresis gels (which is more specific immunologically but is not quantitative).

For the precipitin reaction samples to be assayed, in the form of ribosome—polypeptide complexes, released polypeptides, or soluble proteins, are divided into two 1-ml aliquots; one is treated with 0.05 ml of homologous antiserum and the other with 0.05 ml of heterologous antiserum. Antibody is normally in excess, which does not depress the degree of precipitation. After incubation at 37°C for 1 hr, the tubes are chilled. In order to produce an equal amount of precipitate in both sets of tubes (to equalize the possibility of coprecipitation of nonspecific labeled material), and also to detect any soluble specific antigen—antibody complexes, and excess of sheep anti-rabbit serum is added to all tubes. This technique should yield precipitates sedimenting equal amounts of nonspecific radioactivity and provide a large enough precipitate with which to work. After incubation for a further 30 min at 37°C, the precipitates are sedimented by centrifugation at 4000 × g for 30 min. The pellets are surface-rinsed, resuspended in 2 ml of 0.9% saline using a vortex mixer, and recentrifuged. The pellet is finally resuspended in 0.1 ml of saline and, after treatment with 2 N NaOH, the protein is precipitated with 10% TCA, washed, and the radioactivity is counted.

For further analysis of serologically precipitated, released nascent polypeptides, the antibody—antigen complexes are solubilized in 8 M urea and 0.05 M 2-mercaptoethanol. Molecular size of the labeled polypeptides can be estimated by acrylamide gel electrophoresis [6].

All measurements of radioactivity precipitated with homologous antiserum are corrected for nonspecific precipitation by subtracting the amount of radioactivity precipitated in the corresponding controls (see Figure 5A). The degree of nonspecific precipitation of radioactivity varies with different types of subcellular preparation. This effect is presumably a result of the differing stabilities of the various preparations as well as the different amounts of cross-reacting material present. For example, when ribosome preparations are treated with antiserum to assay the amount of attached nascent antigenic protein, it is necessary to maintain them in dialysis buffer since spontaneous precipitation occurs in saline. Also, the time of treatment should be kept to a minimum, that is, all reactions prior to protein precipitation should be performed on the same day. Nonspecific precipitation varies from 20 to 100% of specific precipitation values.

For assay of specific incorporation by immunoelectrophoresis and autoradiography, the labeled samples must first be reduced in volume, This may be done either by concentrating the sample by dialysis against polyethylene glycol (molecular weight = 6000) or by pelleting the ribosomes by centrifugation at 110,000 χ g for 90 min and releasing nascent polypeptides by treating the pellet with 0.5% Brij 58 and 0.02 M EDTA in a total volume of 0.05 ml. In order to obtain visible precipitin arcs, and also to provide a built-in control, 0.05 ml of a mixture of homologous and heterologous antigenic protein (both at a concentration of 1 mg/ml) is added to the sample as a carrier. For electrophoresis, microscope slides are coated to a depth of 2 mm with 1% Noble agar (Difco) in Veronal acetate buffer (I = 0.05, pH 8.6). Fifty-microliter samples are applied to wells in the agar and electrophoresis is carried out in Veronal acetate buffer

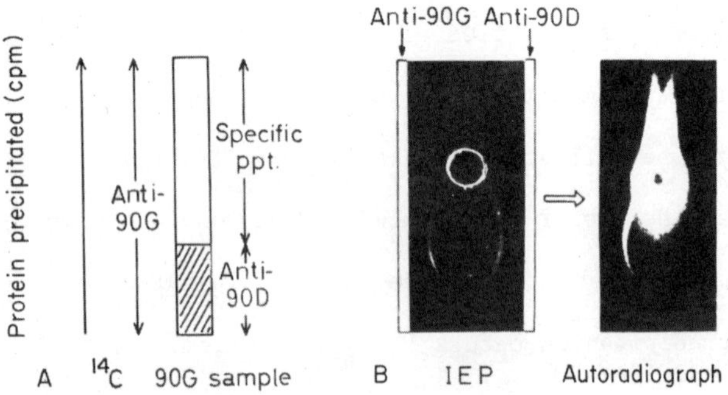

Fig. 5. Techniques for demonstrating radioactive labeling of specific antigenic proteins. (A) Precipitin reaction with a labeled ribosomal fraction derived from 90G cells. The sample is divided into two parts, one treated with homologous (anti-90G) serum, the other with heterologous (anti-90D) serum (see text). The specific reaction is calculated by subtraction. (B) immunoelectrophoresis arcs of carrier homologous (90G) and heterologous (90D) antigens plus EDTA-released labeled polypeptides derived from 90G cells. Autoradiograph shows only homologous antigen protein labeled.

(I = 0.5, pH 8.6) for 4 hr at 20 volts/cm. After electrophoresis homologous and heterologous antiserum are applied on longitudinal filter paper strips between the sample wells (see Figure 5B). Precipitin bands of the homologous and heterologous antibody—antigen reactions should develop by double diffusion after 1—2 days; the gels are kept in a moist atmosphere. After the gels are thoroughly washed to remove nonprecipitated material, the precipitin arcs are autoradiographed by exposing the slides supporting the dried agar gels to Royal Blue x-ray film (Kodak). The surface of the agar and the film should be held firmly in contact between two thick glass plates. Exposure time is generally from 3 to 8 weeks, depending upon the specific activity of the sample. Manipulations and development of the film should take place in total darkness. All exposed films are developed for 4 min at 20°C in DK-80 developer (Kodak). A typical autoradiograph of released antigenic polypeptides is shown in Figure 5B.

V. METHODS OF TRACING THE BIOSYNTHETIC PATHWAY OF SURFACE ANTIGEN PROTEINS

A. Synthesis by Isolated Gradient Fractions

By use of the *Paramecium* cell-free system, in conjunction with the separation and assay techniques described in the previous sections, the mechanism of biosynthesis of the surface antigens can be studied.

For instance, it is found that surface antigens can be synthesized by isolated membrane-bound ribosomes, yet a certain amount of activity in synthesizing antigen polypeptides is found in free ribosome fractions [6]. The polyribosomes active in antigen synthesis have discrete sedimentation values of between 240 and 300 S. A limited range in sedimentation coefficient indicates first that the polyribosomes in question are not random breakdown products, and second that the size of polypeptides synthesized by such units can be estimated. In fact, these S values correlate well with the size of polyribosome that would be required to form a subunit (molecular weight = 100,000) of the antigen. However, since each subunit is composed of three polypeptides (molecular weight = 35,000), the sedimentation value may indicate an aggregate of polyribosomes held together by virtue of their nascent polypeptide chains. An alternative explanation is the presence of a polycistronic mRNA. That *in vitro* aggregation of labeled antigen polypeptides does in fact occur can be shown by releasing the polypeptides with puromycin. After 10 min of incubation at 27°C, puromycin is added to a concentration of 200 μg/ml and the incorporation mixes are incubated for a further 10 min. After centrifugation at 110,000 × g for 90 min, the supernatant is layered on a 20-ml, 5—20% sucrose gradient and centrifuged at 30,000 rpm for 30 hr. The specifically precipitated radioactive protein is found to sediment almost as fast as the native protein antigen marker (8 S) [4]. However, under reducing conditions (8 M urea, 0.05 M 2-mercaptoethanol), serologically precipitated nascent polypeptides mainly migrate slightly further through 8% acrylamide gels than do reduced antigen markers, indicating that the nascent peptides are incompletely formed single polypeptides [6].

Evidence for "precursor" polyribosomes synthesizing such single polypeptides (i.e., using monocistronic mRNA) can be provided by the use of antiserum against reduced carboxymethylated (RCM) surface antigen. The protein is reduced with 0.12 M 2-mercaptoethanol in 8 M urea and alkylated with iodoacetic acid [9]. The RCM antigen is recovered by dialysis against several changes of 5 mM $(NH_4)_2CO_3$ and lyophilized. Antiserum is prepared as described previously. Anti-RCM serum specifically precipitates polyribosomes smaller in size than does the antiserum against the native protein. Also, the puromycin-released antigenic polypeptides have a low sedimentation value (<2 S) which indicates that they lack the ability to aggregate. The size of the anti-RCM antigen-precipitated ribosomes varies from 150 to 200 S and corresponds well with the value expected of polyribosomes synthesizing a protein of approximate molecular weight 30,000, which is that of a single antigen protein polypeptide.

This evidence for a possible initial stage of biosynthesis of single polypeptides with later aggregation of their polyribosomes by virtue of their nascent polypeptide chains has received added support from another valuable approach—the use of conjugated antibodies.

B. Location of Nascent Antigen by the Use of Labeled Antibodies

^{125}I-labeled antibodies can be prepared by the conjugation procedure described by Johnson *et al.* [23]; each 1 ml of IgG sample (containing 20—25 mg of protein) is treated with 160 μCi of $Na^{125}I$[24]. When mixed with freshly extracted ribosomes and eluted from a column of Sepharose 4B (Pharmacia) with 0.5 M tris—HCl (pH 7.6), ^{125}I-conjugated homologous antibodies are bound only to large aggregates of ribosomes [10]. Labeled heterologous antibodies are not bound. The specific binding of ^{125}I-globulins to nascent polypeptides in the polyribosome fractions can be further investigated by gradient analysis. The ^{125}I binds to polyribosomes, some of which precipitate and pellet to the bottom of the gradient. However, peaks of radioactivity are found in soluble complexes sedimenting at 210 and 230 S.

Thus by two different techniques two classes of ribosomes can be shown to be involved in antigen biosynthesis and may represent temporal phases of the process.

C. Membrane—Ribosome Associations

That association of antigen-synthesizing ribosomes with membranes is attributable to chemical attachment, rather than to cosedimentation of very large ribosomal aggregates, can be demonstrated by three methods: (1) Ribosomes with associated labeled antigen readily adsorb to washed membrane preparations, whereas soluble labeled antigen does not adsorb to membranes. (2) Less than 10% of labeled antigen associated with membrane-bound ribosomes sediments through 2 M sucrose on prolonged centrifugation (130,000 × g for 5 hr) whereas 85% of the radioactivity associated with free ribosomes sediments under the same conditions. (3) Treatment of labeled membrane fractions with 0.5% sodium DOC does not release any very large ribosomal complexes. In fact, the largest size of released polyribosome active in antigen protein synthesis is about 300 S.

The nature of attachment of antigen-synthesizing polypeptides to membranes appears not to involve carbohydrate moieties (which play an important role in the transport of immunoglobulins [25]), since no labeled antigen is serologically precipitated from cells labeled with [^{3}H]galactose or [^{3}H]glucosamine.

D. Transport and Cell Surface Appearance of Antigen Protein

The means by which the antigen protein is transported from the rough membrane (endoplasmic reticulum) to the cell surface is really a topographical problem and as such is outside the scope of this chapter. Briefly, antigen transport can be followed in cell sections by locating electron-dense antibody conjugates in electron micrographs or by locating radioactively labeled antibody by high-resolution autoradiography. A starting point in the process of biosynthesis of paramecium antigen is provided by changing the environmental conditions to stimulate the formation of a new type antigen, a process called serotype transformation. The new type antigens on the cell surface can be accurately located by means of ferritin-conjugated antibodies. Ferritin (six times crystallized, Calbiochem) is conjugated to IgG with the bifunctional coupling agent m-xylylene diisocyanate (Kodak) [26]. The new type antigen first appears at the base of the cilia and later spreads to cover the surface of the pellicle and the length of the cilia. That the ferritin reaction detects newly synthesized antigen can be demonstrated by correlating the appearance of ^{35}S-labeled protein on surface structures with the sites of ferritin reaction. Localization of conjugated antibodies in sections of the cytoplasm is difficult, but the consensus of opinion is that the antigen is transported from the cisternal spaces of the endoplasmic reticulum via "Golgi-like" vesicles which fuse to specific sites, the parasomal sacs, which open to the exterior of the cell at the base of the cilia [24].

VI. OTHER PROTOZOAN SYSTEMS

With the *Paramecium* surface antigen system we have been fortunate in having at our disposal an extensive genetic, chemical and cytological background that has been prepared over a number of years (reviewed recently [17]). Now, other potential systems are beginning to mature. The filamentous trichocyst protein of *Paramecium* has been chemically characterized [27] as has the fibrous, actinlike ciliary protein of *Tetrahymena* [28]. Microtubular proteins differentiate extensive and elaborate organelles in the ciliate *Nassula* [29], as well as in other protozoa, and their biosynthesis is well worth studying. Recently, the serological and cytological aspects of the surface antigens of the flagellated protozoa *Trypanosoma* have been studied [30]; the basic interest is much the same as that in *Paramecium*, namely, the control mechanisms involved in antigen expression.

ACKNOWLEDGMENTS

I should like to take this opportunity to thank all those who helped me during my formative years in the Institute of Animal Genetics, University of Edinburgh, particularly Professor G. H. Beale, F.R.S., Dr. J. O. Bishop, and Dr. I. G. Jones. I have quoted freely from the work of my former collaborator Dr. R. E. Sinden, to whom I am most grateful. Finally, I thank Mrs. A. R. Oxbrow without whose technical assistance little would have been achieved.

REFERENCES

[1]. J. Magar and F. Lipmann, *Proc. Natl. Acad. Sci. U.S.*, **44**, 305 (1958).

[2]. A. H. Reisner and H. Macindoe, *J. Gen. Microbiol.*, **47**, 1 (1967).

[3]. J. Sommerville, Ph.D. Thesis, Univ. of Edinburgh, Edinburgh, 1967.

[4]. J. Sommerville, *Biochim. Biophys. Acta*, **149**, 625 (1967).

[5]. J. Sommerville and R. E. Sinden, *J. Protozool.*, **15**, 664 (1968).

[6]. J. Sommerville, *Biochim. Biophys. Acta*, **209**, 240 (1970).

[7]. T. M. Sonneborn, in Methods in Cell Physiology (D.M. Prescott, ed.), Vol. 4, Academic Press, New York, 1970), p. 241.

[8]. A. T. Soldo and W. J. Van Wagtendonk, J. Protozool., 16, 500, (1969).

[9]. I. G. Jones, Biochem. J., 96, 17 (1965).

[10]. R. E. Sinden, J. Protozool., 17A, 24 (1970).

[11]. J. K. Chesters, Biochim. Biophys. Acta, 114, 385 (1966).

[12]. A. H. Reisner, J. Rowe, and H. M. Macindoe, J. Mol. Biol., 32, 587 (1968).

[13]. U. Loening, J. Mol. Biol., 38, 355 (1968).

[14]. J. R. Preer, J. Immunol., 83, 378 (1959).

[15]. J. O. Bishop, J. Gen. Microbiol., 30, 271 (1963).

[16]. J. O. Bishop, Biochim. Biophys. Acta, 50, 471 (1961).

[17]. J. Sommerville, Advan. Microbial Physiol., 4, 131 (1970).

[18]. A. H. Reisner, J. Rowe, and H. M. Macindoe, Biochim. Biophys. Acta, 188, 196 (1969).

[19]. E. Balbinder and J. R. Preer, J. Gen. Microbiol., 21, 156 (1959).

[20]. I. Finger, F. Onorato, C. Heller, and H. B. Wilcox, J. Mol. Biol., 17, 86 (1966).

[21]. W. J. Herbert, in Handbook of Experimental Immunology, D.M. Weir, ed.), Blackwell, Oxford, (1967), p. 1207.

[22]. R. E. Sinden, J. Protozool., 17, 600 (1970).

[23]. H. M. Johnson, L. Day, and D. Pressman, J. Immunol., 84, 213 (1960).

[24]. R. E. Sinden, Ph.D. Thesis, Univ. of Edinburgh, Edinburgh, (1969).

[25]. J. W. Uhr and I. Schenkein, *Proc. Natl. Acad. Sci. U.S.* *66*, 952 (1970).

[26]. K. C. Hsu, in *Methods in Immunology and Immunochemistry* (C. A. Williams and M. W. Chase,eds.), Academic Press, New York, (1967), p. 397.

[27]. E. Steers, J. Beisson, and V. T. Marchesi, *Exptl. Cell Res.*, *57*, 392 (1969).

[28]. F. L. Renaud, A. J. Rowe, and I. R. Gibbons, *J. Cell Biol.*, *36*, 79 (1968).

[29]. J. B. Tucker, *J. Cell Sci.*, *7*, 793 (1970).

[30]. K. Vickerman and A. G. Luckins, *Nature*, *224*, 1125 (1969).

Chapter 8

PREPARATION AND ASSAY OF HEMOGLOBIN mRNA

Jerry B. Lingrel

Department of Biological Chemistry
College of Medicine, University of Cincinnati
Cincinnati, Ohio

I. INTRODUCTION 231

II. ISOLATION OF HEMOGLOBIN mRNA 233

A. Induction of Reticulocytosis 234

B. Preparation of Ribosomes 236

C. Preparation of Hemoglobin mRNA from Ribosomes. . 238

III. ASSAY OF HEMOGLOBIN mRNA 247

A. General Considerations 247

B. Reticulocyte Cell-Free System. 248

C. Analysis of Products Synthesized in the Cell-Free System 252

REFERENCES 261

I. INTRODUCTION

Jacob and Monod [1] first used the term messenger RNA (mRNA) in 1961 in their unifying interpretation of experiments concerning the synthesis of RNA after bacteriophage infection of Escherichia coli and the kinetics of enzyme induction and repression. Although much has been learned concerning the function and properties of this RNA, only recently have active mRNAs been isolated from organisms other than RNA viruses.

Using indirect criteria, Marbaix and Burny [2], Burny and Marbaix [3], Huez et al. [4], and Chantrenne et al. [5] isolated an RNA from rabbit reticulocyte polysomes that exhibited many of the properties expected for hemoglobin mRNA. Evans and Lingrel [6] isolated a similar RNA from mouse reticulocyte ribosomes and presented additional indirect evidence that the 9 S RNA was hemoglobin mRNA [7].

Proof that the 9 S RNA was hemoglobin mRNA was based on the absolute criterion for a mRNA, namely, its ability to direct the synthesis of a specific protein. Using this criterion, Laycock and Hunt [8] showed that rabbit 9 S RNA directed the synthesis of rabbit hemoglobin when added to an E. coli cell-free protein-synthesizing system, and Lockard and Lingrel [9] showed that mouse reticulocyte 9 S RNA directed the synthesis of mouse globin β chains when added to a rabbit reticulocyte cell-free system. Schapira et al. [10] showed that an RNA fraction isolated from reticulocytes was responsible for the amino acid ambiguities observed in the hemoglobin of different rabbits. Recently, mouse 9 S RNA has been shown to direct the synthesis of both mouse α and β chains when added to either a guinea pig (R. Jones and J. B.

Lingrel, unpublished) or duck (R. E. Lockard and J. B. Lingrel, unpublished) reticulocyte cell-free system. Also, rabbit 9 S RNA has been shown to direct the synthesis of rabbit globin in a guinea pig reticulocyte system (R. Jones and J. B. Lingrel, unpublished).

Myosin mRNA has been isolated from chick embryonic muscle by Heywood and Nwagwu [11], utilizing the observation that this protein is made on large polyribosomes. It has not been extensively purified. RNAs tentatively identified as histone mRNAs were isolated from cleaving embryos of *Arbacia punctulata* eggs by Kedes and Gross [12] and from S-phase HeLa cells by Borum et al. [13] and Gallwitz and Mueller [14].

II. ISOLATION OF HEMOGLOBIN mRNA

Reticulocytes, which synthesize predominantly hemoglobin [15], provide a good source for the isolation of a specific mRNA because a high percentage of the mRNA of these cells is specific for hemoglobin.

The general procedure for preparing hemoglobin mRNA involves isolating ribosomes from reticulocytes, incubating them with sodium dodecyl sulfate (SDS) to dissociate RNA and protein, and purifying the mRNA by sucrose density gradient centrifugation. The assay is based on the definition of mRNA, that is, its ability to direct the synthesis of hemoglobin, or at least globin α and β chains. It employs a cross system in which the hemoglobin mRNA of one species is added to a reticulocyte cell-free system prepared from another, and the hemoglobin or globin chains synthesized under the direction of the added mRNA are determined.

A. Induction of Reticulocytosis

Hemoglobin synthesis occurs in reticulocytes and earlier erythroid precursor cells but not in mature erythrocytes. In the normal animal, reticulocytes comprise only a few percent (1-5%) of the circulating cells, whereas earlier cells are almost never present. The number of reticulocytes in the circulation can be greatly enhanced by inducing anemia either by bleeding or with hemolytic agents such as phenylhydrazine. Both procedures appear to be equally effective, but phenylhydrazine-induced anemia is much easier to use experimentally and is the only procedure described in this chapter.

Two solutions of phenylhydrazine are used. One is a 2.5% solution made by dissolving 2.5 g of phenylhydrazine-HCl and 2.17 g of sodium acetate in approximately 75 ml of water, adjusting the pH to 7.0 with 5 N NaOH, and adding water to 100 ml. The second is an 0.8% phenylhydrazine solution made in the same way, except that 0.8 g of phenylhydrazine is used. The presence of sodium acetate in these solutions is based upon historical reasons, and its omission does not appear to reduce the reticulocytosis. The 2.5% phenylhydrazine solution is used with rabbits, guinea pigs, and ducks, while the 0.8% solution is used with mice.

Reticulocytosis in New Zealand white rabbits is produced by daily subcutaneous injections of 2.5% phenylhydrazine at a dose of 0.25 ml/kg for 6 consecutive days. The animals are not given an injection on the seventh day. They are bled by cardiac puncture on the eighth day with the use of a 50-ml syringe containing 750 units of heparin. The blood is immediately cooled to 2°C. Rabbits weighing 2.5 kg have been found to be most satisfactory; they exhibit good reticulocytosis with low mortality. Blood obtained from these animals has a hematocrit in the range 15-23%, and greater than 90% of the cells are reticulocytes.

Reticulocyte counts are made on smears prepared from blood that has been mixed with an equal volume of 0.48% new methylene blue in 87 mM potassium oxalate and allowed to stand for 10 min. Duck blood is stained by adding 2 vol of the new methylene blue solution and waiting 30 min before making the smear.

Reticulocytosis in guinea pigs is produced by six consecutive daily injections of 2.5% phenylhydrazine (0.48 ml/kg) [16]. On the seventh day the animals are anesthetized with ether, and the blood is removed by heart puncture with a heparinized syringe. The reticulocyte count of anemic guinea pig blood is usually lower that that obtained for rabbits.

Mice (25- to 30-g Swiss Cox) are made anemic by daily intramuscular injections of 0.1 ml of 0.8% phenylhydrazine for 6 days and bled on the seventh day [17]. The mice are anesthetized with 5 mg of sodium pentobarbital and bled by opening the chest cavity, adding 100 units of heparin to the cavity, and collecting the blood from the cavity after cutting one or more of the blood vessels surrounding the heart. Reticulocyte counts of approximately 90% are obtained. Reticulocytosis in ducks (2.2- to 2.8-kg Pekin ducks) is induced by daily intramuscular injections of 2.5% phenylhydrazine at a dose of 0.4 ml/kg for 6 days, and the animals are bled by heart puncture on the eighth day. Approximately 90% of the cells obtained are reticulocytes.

B. Preparation of Ribosomes

The blood from anemic animals is strained through three layers of cheesecloth to remove any clots that may be present, and the cells are collected from the plasma by centrifugation at 500 X g for 10 min at 2°C. All subsequent operations are performed at 2°C; these are summarized in Table I. The cells are washed four times with an isotonic saline solution (solution I) which contains 130 mM NaCl, 7.4 mM $MgCl_2$, and 5 mM KCl [18]. Four volumes of solution I are used in each washing, and the cells are centrifuged at 500 X g for 10 min. After the first or second wash, the thin layer of white cells above the

TABLE I

Preparation of Ribosomes

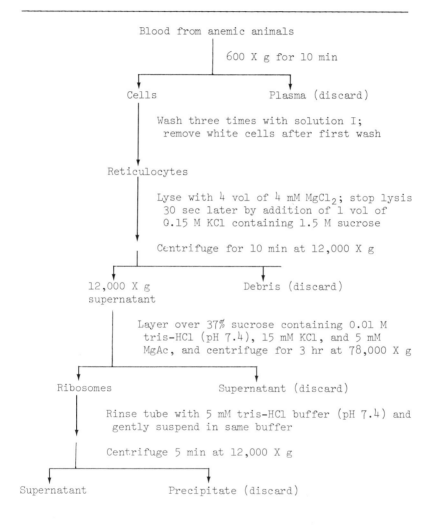

Blood from anemic animals

↓ 600 X g for 10 min

Cells → Plasma (discard)

Cells

↓ Wash three times with solution I; remove white cells after first wash

Reticulocytes

↓ Lyse with 4 vol of 4 mM $MgCl_2$; stop lysis 30 sec later by addition of 1 vol of 0.15 M KCl containing 1.5 M sucrose

↓ Centrifuge for 10 min at 12,000 X g

12,000 X g supernatant → Debris (discard)

12,000 X g supernatant

↓ Layer over 37% sucrose containing 0.01 M tris-HCl (pH 7.4), 15 mM KCl, and 5 mM MgAc, and centrifuge for 3 hr at 78,000 X g

Ribosomes → Supernatant (discard)

Ribosomes

↓ Rinse tube with 5 mM tris-HCl buffer (pH 7.4) and gently suspend in same buffer

↓ Centrifuge 5 min at 12,000 X g

Supernatant → Precipitate (discard)

Adjust ribosome concentration to 8 mg/ml and store at -60°C.

---

reticulocytes becomes quite visible and is removed by aspiration. The packed cells are lysed by adding 4 vol of 4 mM $MgCl_2$ with gentle stirring; the lysis is stopped after 30 sec by making the solution isotonic with 1 vol (equal to the volume of the cells) of a solution containing 0.15 M KCl and 1.5 M sucrose. The 30-sec lysis is sufficient to lyse essentially all the reticulocytes, while few if any of the remaining white cells are broken. After lysis the unbroken cells, cell debris, and mitochondria are removed by centrifugation at 12,000 X g for 10 min. The supernatant solution is layered over 7 ml of 36% (w/w) sucrose containing 10 mM tris-HCl (pH 7.4), 15 mM KCl, and 5 mM magnesium acetate in a 30-ml centrifuge tube and spun for 3 hr at 78,000 X g [4]. Hemoglobin and other proteins are removed from the ribosomes as they pass through the sucrose layer. The supernatant solution is discarded and the inside of the tube, as well as the straw-colored pellet, is rinsed with 5 mM tris-HCl (pH 7.4); the pellet is then suspended with gentle homogenization in the same buffer. The solution is centrifuged for 5 min at 12,000 X g, and the supernatant solution is adjusted to a ribosomal concentration of 8 mg/ml (using an extinction coefficient of 10 A_{260} units/mg). The ribosome solution can be kept frozen at -60°C for 1 year or longer without loss of mRNA activity.

C. Preparation of Hemoglobin mRNA from Ribosomes

Several procedures have been described for the isolation of hemoglobin mRNA from ribosomes.

Burny and Marbaix [3] isolated the mRNA by sucrose density gradient centrifugation of SDS-dissociated ribosomes. This procedure involves a minimal amount of handling but requires a zonal rotor for the purification of milligram quantities of mRNA [10]. Laycock and Hunt [8] prepared hemoglobin mRNA by differentially precipitating it, along with another RNA, from total rRNA by a salt fractionation procedure. The other RNA can be removed from the mRNA by sucrose density gradient centrifugation. Temmerman and Lebleu [20] and Labrie [21] prepared mRNA from a mRNA-protein particle obtained from ribosomes. The particle is librated from ribosomes by treatment with disodium EDTA and is subsequently separated from the ribosomal subunits by sucrose density gradient centrifugation. The mRNA is prepared from the particle by sucrose density gradient centrifugation of the SDS-treated particle. Alternatively, the mRNA can be isolated from the particle by the SDS-phenol procedure. mRNA can also be isolated by polyacrylamide gel electrophoresis of total rRNA [22].

Isolation of hemoglobin mRNA by sucrose density gradient centrifugation of SDS-treated ribosomes has been routinely used by our laboratory to prepare highly purified functional mRNA. This reproducible and relatively simple method, which is summarized in Table II, is the only isolation procedure discussed in this chapter. mRNA (0.6-2 mg) can be prepared by a single centrifugation using a zonal rotor [19], whereas smaller quantities are isolated using conventional swinging-

TABLE II

Preparation of Hemoglobin mRNA

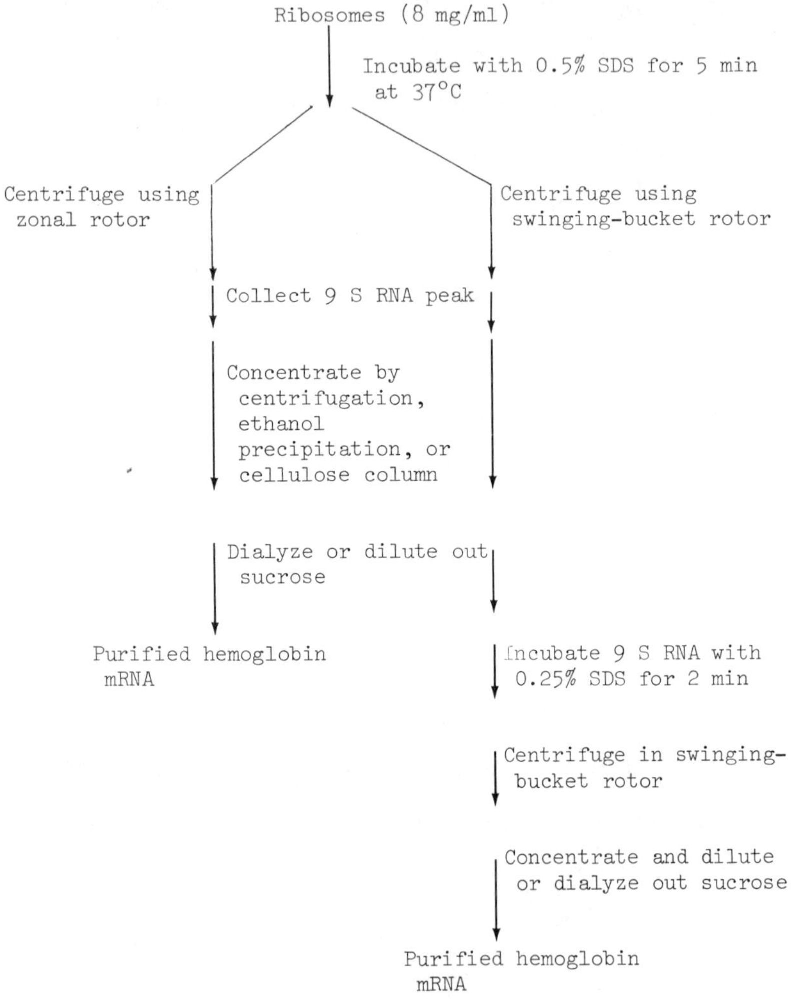

bucket rotors [3,9]. In either case the ribosomes [8 mg/ml in 5 mM tris-HCl (pH 7.4)] are dissociated into RNA and protein by the addition of 1/20 vol of 10% SDS in H_2O followed by a 5 min incubation at 37°C. The dissociated ribosomes are either immediately layered on the sucrose gradient or cooled in ice for a short period of time (up to 20 min) before layering. For swinging-bucket rotors a 5-20% linear sucrose density gradient is employed. The gradients are prepared with RNase-free sucrose obtained from Mann Research Laboratories, and all solution are prepared on a w/w basis in 5 mM tris-HCl (pH 7.4). In a typical run using an SW-27 rotor, 8 mg of ribosomes are layered on each gradient and centrifuged at 27,000 rpm for 26 hr at 2°C. The RNA pattern from such a centrifugation is shown in Figure 1A. With an SW-40 rotor 2 mg of ribosomes are layered on the gradient, and centrifugation is for 14 hr at 40,000 rpm.

The peak fractions containing the mRNA are pooled and concentrated by either centrifugation or ethanol precipitation. Concentration by centrifugation is performed by spinning the pooled fractions for 20 hr at 105,000 X g in a Beckman Type 40 fixed-angle rotor. At the end of the spin, more than 90% of the RNA is located in the bottom 1 ml. Some of the RNA occurs as a pellet, but most of it is found in solution just above the pellet. Because a second sucrose gradient centrifugation is

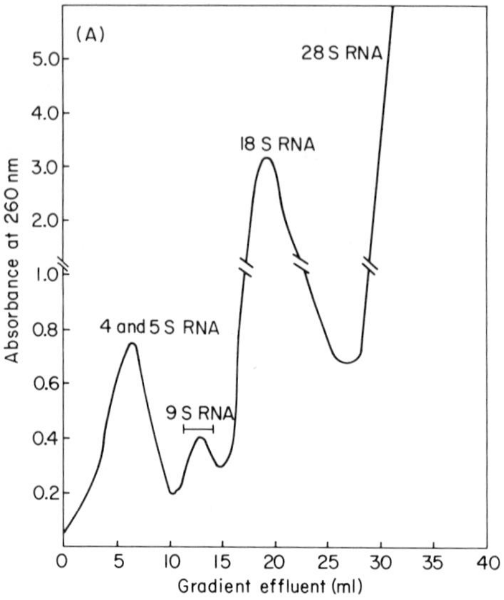

Fig. 1. Preparation of mouse reticulocyte mRNA using a swinging-bucket rotor. Centrifugation was performed in a Spinco SW-27 rotor at 27,000 rpm for 26 hr at 1.8°C. Thirty-seven-milliliter 5-20% linear sucrose gradients were used. At the end of the run, the gradients were analyzed by pumping 40% sucrose into the bottom of the tube and continuously recording the absorbance at 260 nm of the solution emerging from the top as it passed through a 0.2-ml flow cell with a 1-cm light path. (A) Centrifugation of SDS-treated ribosomes. Eight milligrams of ribosomes in 1 ml of 5 mM tris-HCl (pH 7.4) were incubated for 5 min at 37°C with 0.05 ml of 10% SDS before layering on the gradient. (B) Recentrifugation of 9 S

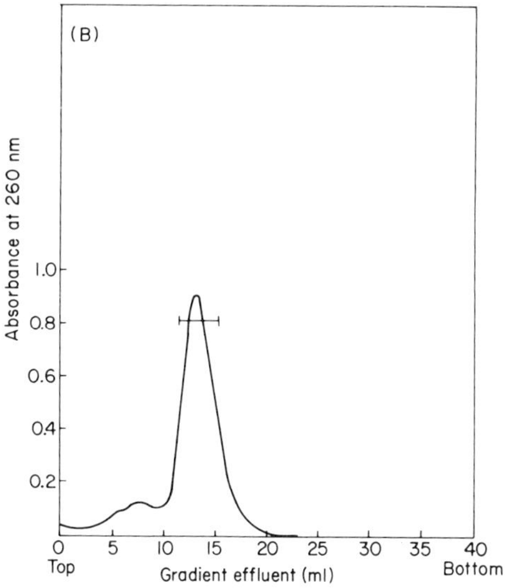

RNA. Fractions pooled from the 9 S RNA region, as indicated in Figure 1A, were concentrated by centrifugation; the sucrose concentration was lowered to below 5% by diluting with 11 vol of 5 mM tris-HCl (pH 7.4), and the RNA was reconcentrated by centrifugation. Three milliliters, containing 200 μg of RNA, were incubated for 2 min at 37°C with 0.25% SDS and layered on the gradient. (C) Analysis of purified mRNA. The peak fractions, as indicated in Figure 1B, were pooled; the RNA was concentrated and the sucrose diluted as above. Three milliliters of this solution, containing 200 μg of 9 S RNA, were incubated for 2 min at 25°C with 0.25% SDS and layered on the gradient.

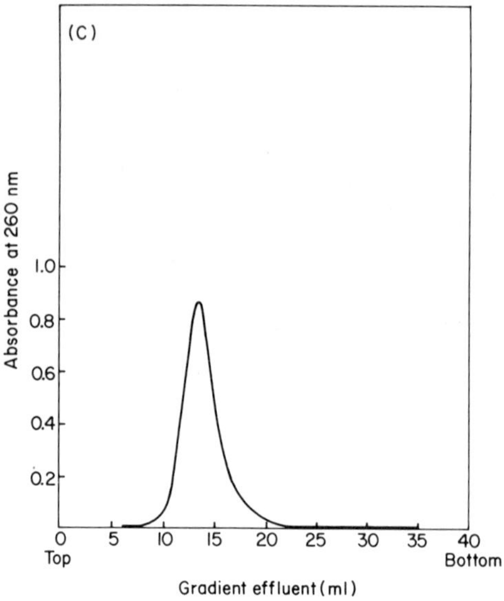

required to purify the mRNA, the sucrose concentration must be lowered to less than 5% before it can be layered on a second 5-20% gradient. This is accomplished by adding 11 ml of 5 mM tris-HCl buffer (pH 7.4) to the 1 ml remaining at the bottom of each tube and centrifuging again for 20 hr at 105,000 X g. Again, the RNA is concentrated in the bottom 1 ml of the tube and in the small pellet. The pellet is resuspended in the bottom 1 ml of each tube, incubated for 2 min at 37°C with 0.25% SDS to dissociate any contaminating protein from the RNA, and recentrifuged on a 5-20% sucrose gradient.

When the mRNA is concentrated by ethanol precipitation, solid NaCl is added to the pooled fractions to give a final concentration of 0.25 M; then 2 vol of 95% ethanol at a temperature of -20°C are added. The mixture is allowed to stand overnight at -20°C, and the precipitate is collected by centrifugation at 15,000 X g for 15 min. The RNA is dissolved in a minimal amount of 5 mM tris-HCl buffer (pH 7.4). The concentrated RNA is made 0.25% in SDS and warmed to 37°C for 2 min. Three to four milliliters of the solution, concentrated by centrifugation or ethanol precipitation, are layered on each 35-ml 5-20% gradient (SW-27 rotor) and centrifuged for 26 hr at 27,000 rpm. Figure 1B shows the RNA pattern from such a centrifugation. The tubes indicated by the bar are pooled and concentrated by either the centrifugation or the ethanol precipitation method. A typical density gradient analysis of the pooled fractions is shown in Figure 1C. As can be seen, the mRNA migrates as a single band. Recovery of the mRNA from ribosomes is 50% using this procedure. Sucrose depresses mRNA-directed incorporation activity, so its concentration is reduced as outlined above before the mRNA is added to the cell-free system. When ethanol is used for concentration, it is removed from the preparation by dialysis against several changes of 5 mM tris-HCl (pH 7.4). The dialysis tubing is boiled in 5 mM EDTA for 1 hr and washed with distilled water before using.

For the preparation of larger quantities of hemoglobin mRNA, a Beckman Ti-15 zonal rotor is used (H. Burr and J. B. Lingrel, unpublished). The gradient in this rotor is convex and is formed by using a constant-volume mixing flask containing 400 ml of 5% sucrose and a reservoir, feeding into the mixing flask, which contains 20% sucrose. The gradient is formed by pumping the solution from the mixing flask into the rotor. Two-hundred milligrams of SDS-dissociated ribosomes (25 ml) are layered on the gradient and overlayed with 50 ml of 5 mM tris-HCl buffer (pH 7.4). The rotor is run at 35,000 rpm for 36 hr at 2°C. Figure 2 shows the RNA pattern from such a run. The peak fractions are pooled and the mRNA is concentrated by one of the methods described above. The mRNA can also be concentrated by adsorption to cellulose [19]. The pooled sucrose gradient fractions that contain 9 S RNA (200 ml) are made 0.1 M in NaCl and 0.5% with respect to SDS. After the addition of 0.54 vol of ethanol, the solution is passed through a 2 X 5 cm cellulose column. The mRNA is eluted with 10-15 ml of water.

The mRNA fraction obtained from the zonal rotor gives a pattern similar to that shown in Figure 1C and is not purified further. Both mRNA preparations, when electrophoresed in 5% acrylamide gels, give a major band which comprises more than 90% of the RNA and one or two small bands (H. Burr and J. B. Lingrel, unpublished). The mRNA preparations are stored at -60°C.

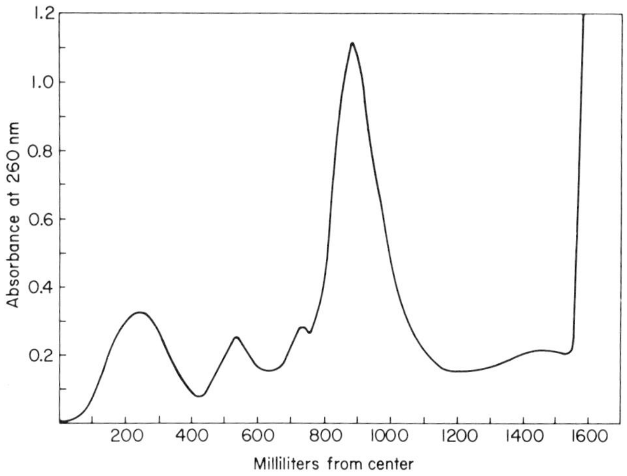

Fig. 2. Isolation of mRNA using a zonal rotor. Two-hundred milligrams of rabbit ribosomes in 25 ml of 5 mM tris-HCl (pH 7.4) were incubated for 5 min at 37°C with 1.25 ml of 10% SDS and immediately pumped into a Beckman Type 15 zonal rotor containing the sucrose gradient. The sample was overlayed with 50 ml of 5 mM tris-HCl (pH 7.4) and centrifugation was for 36 hr at 35,000 rpm. The temperature was maintained at 2°C. The 260-nm absorbance was continuously recorded as the solution was pumped out of the rotor.

III. ASSAY OF HEMOGLOBIN mRNA

A. General Considerations

The hemoglobin mRNA assay is based on the ability of RNA to direct the synthesis of hemoglobin α and β chains. This is accomplished by adding mRNA to a reticulocyte cell-free system and determining the amount of α and β chains made.

The choice of cell-free system used in the assay is based on the ability to separate the α and β chains made under the direction of the added mRNA from those synthesized by the endogenous mRNA of the cell-free system. It should be noted that no attempt is made to lower endogenous mRNA activity.

Mouse mRNA can be assayed in a rabbit, guinea pig, or duck reticulocyte cell-free system, and rabbit mRNA can be assayed in a guinea pig and probably in a duck reticulocyte cell-free system. Once the mRNA has been incubated with the appropriate cell-free system containing radioactive amino acids, globin is prepared and the respective chains are separated by column chromatography and counted. The quantity of label in the newly appearing globin chains is related to the amount of mRNA added.

B. Reticulocyte Cell-Free System

This cell-free system is similar to that described by Adamson et al. [23] and employs a crude reticulocyte lysate rather than a fractionated system. The lysate is used because it is extremely active in hemoglobin synthesis (the activity approaches that of whole cells for at least 10 min) and because it is able to initiate the synthesis of new chains. The latter is a prerequisite for translating added mRNA. The lysate is prepared by adding an equal volume of distilled water containing 5×10^{-5} M hemin at 0°C to the cells. This provides the optimal concentration (determined empirically) of 1 X 10^{-5} M hemin in the cell-free incubation. Hemin is known to protect the cell-free system from inactivation under certain circumstances [23-25]; however, its value in the mRNA assay has not been thoroughly investigated. The cells are gently stirred for 1 min to insure lysis and then centrifuged at 12,000 X g for 10 min to remove cell debris and mitochondria. The supernatant is either used immediately or frozen in liquid nitrogen and stored at -60°C until used. The cell-free system stored in this way does not lose activity for at least 6 months. All operations prior to freezing are performed at 2°C.

Each assay utilizes two cell-free incubations: one contains either a ^{3}H- or ^{14}C-labeled amino acid and mRNA; the other contains the same amino acid labeled with the isotope not used in the first incubation and no mRNA. The latter incubation serves as a control and internal standard.

The cell-free incubations contain, in a final volume of 0.5 ml: 0.20 ml of lysate and 0.05 ml of a solution (solution E) whose composition is given in Table III.

One or more labeled amino acids, usually 4.7 µCi of [^{3}H]leucine and 1 µCi of [^{14}C]leucine (the ratio of ^{3}H to ^{14}C depends upon the relative counting efficiency of these two isotopes), are added to the cell-free system with a commensurate reduction in the unlabeled amino acid so that the final concentration remains unchanged. The mRNA to be

TABLE III

Cell-Free Incubation for the Assay of mRNA

Reticulocytes (0.2 ml)
Solution E (0.05 ml)
Labeled amino acid(s)
mRNA (2-13 μg)
5 mM Tris-HCl (pH 7.4)
Final volume: 0.50 ml

Components of solution E	Concentration		Final concentration in cell-free incubation	
Ammonium acetate	1000	mM	100	mM
Tris-HCl (pH 7.4)	100	mM	10	mM
Magnesium acetate	20	mM	2	mM
ATP	10	mM	1	mM
GTP	2	mM	0.2	mM
Creatine phosphate	150	mM	15	mM
Creatine phosphate kinase	600	μg/ml	60	μg/ml
L-Alanine	0.5	mM	0.05	mM
L-Arginine	0.12	mM	0.012	mM
L-Aspartic acid	0.71	mM	0.071	mM
L-Asparagine	0.5	mM	0.05	mM
L-Cysteine	0.1	mM	0.01	mM
L-Glutamine	1.0	mM	0.10	mM
Glycine	1.3	mM	0.13	mM
L-Histidine	0.6	mM	0.06	mM
L-Isoleucine	0.075	mM	0.0075	mM
L-Leucine	1.0	mM	0.1	mM
L-Lysine	0.45	mM	0.045	mM
L-Methionine	0.085	mM	0.0085	mM
L-Phenylalanine	0.4	mM	0.04	mM
L-Proline	0.35	mM	0.035	mM
L-Serine	0.4	mM	0.04	mM
L-Threonine	0.43	mM	0.043	mM
L-Tryptophan	0.073	mM	0.0073	mM
L-Tyrosine	0.21	mM	0.021	mM
L-Valine	1.0	mM	0.1	mM

assayed (2-13 μg) is added to one of the incubation mixtures, for example, the one containing the [^{3}H]amino acid, and both are made up to 0.5 ml with 5 mM tris-HCl (pH 7.4).

Ammonium acetate provides the monovalent cations required for protein synthesis; it appears to promote greater incorporation of amino acids than potassium ions. Magnesium ions are required for structural integrity of the ribosomes and also for the reactions involving ATP and GTP. As the magnesium concentration is quite low (2 mM), only specific chain initiations are expected. Creatine phosphate and creatine phosphokinase serve to regenerate ATP and GTP from the AMP and GDP formed during synthesis. Adenylic kinase is also required for the regeneration of ATP but is present in the lysate. Although phosphoenolpyruvate and its kinase can serve as an ATP-regenerating source, it is much less effective than creatine phosphate and its kinase [23]. The concentration of each amino acid is based on the earlier work of Borsook et al. [26] with whole cells. Although the ratios of amino acids described are optimal for hemoglobin synthesis in whole cell incubations, this is not necessarily true for the cell-free system; no systematic study has been made, however. Glutamic acid is not present in solution E and is not required by the cell-free system, presumably because of the conversion of glutamine to glutamic acid. The incubation mixtures, with and without mRNA, are incubated at 25°C for 2 hr, cooled to 0°C, and mixed; after 40 mg of the appropriate carrier hemoglobin (mouse hemoglobin if mouse mRNA was added to the cell-free system) is added, globin is prepared according to the method of Rabinovitz and Fisher [27]. The combined cell-free systems containing the carrier hemoglobin are added dropwise to 10 vol of 0.15 N HCl in acetone at -20°C, and the precipitate is collected by centrifugation at this temperature. The resultant precipitate is washed twice with acetone at -20°C and dried in air or under reduced pressure.

C. Analysis of Products Synthesized in the Cell-Free System

Globin chains are separated by column chromatography on Whatman CM-52 carboxymethyl cellulose. The actual conditions vary depending on the type of cross system used and they are discussed separately. For assaying mouse hemoglobin mRNA, a rabbit, guinea pig, or duck cell-free system may be used. Mouse α chains cannot be resolved from rabbit chains, and therefore only β-chain mRNA activity can be assayed in a rabbit cell-free system. Both mouse α and β chains can be separated from guinea pig and duck globin chains, thus making it possible to assay both α- and β-chain mRNA in these systems.

Globin chains are separated on Whatman CM-52 carboxymethyl [illegible]lose using sodium phosphate gradients, or a combination of sodium phosphate and pH gradients. Sodium phosphate buffers are made by dissolving $Na_2HPO_4 \cdot 7H_2O$ to give the indicated molarity and adjusting the pH with concentrated phosphoric acid. The buffers used to equilibrate and elute the columns contain 8 M urea and 50 mM β-mercaptoethanol. Urea dissociates the globin chains and mercaptoethanol protects them from oxidation.

Whatman CM-52 contains many fine particles which severely restrict the flow of urea solutions and these particles are removed by decantation prior to pouring the column. Eighty grams of CM-52 are suspended in 1 liter of starting buffer, without urea and mercaptoethanol, and placed in a 1-liter graduated cylinder. All carboxymethyl cellulose that does not settle to the bottom in 25 min is discarded. This procedure is repeated twice. The CM-52 that has been freed of fine particles is mixed with 0.5 vol of starting buffer, and a column 1.6 X 15 cm is poured. The column is equilibrated with 30 ml of starting buffer.

Analytical grade urea contains some ions and ultraviolet-absorbing material. These are removed from the urea solution, prior to adding sodium phosphate and mercaptoethanol, by adding 75 g of the ion exchanger AG-501 X 8 (D) (20- to 50-mesh), a mixed-bed resin, to each liter of 8 M urea and stirring for 1-2 hr. The urea may be dissolved and deionized simultaneously. The mixed-bed resin is removed by filtration and can be reused. Columns are run at room temperature with a flow rate of 30 ml/hr, and 5- to 10-ml fractions are collected.

In order to determine the radioactivity of the column eluate, 1 drop of 2% serum albumin as a carrier is added to each tube, and the protein is precipitated with a volume of 20% trichloroacetic acid (TCA) equal to that of the fraction collected. The final concentration of urea must be below 5 M to insure quantitative precipitation of the protein. The precipitate is collected at room temperature on 24-mm glass fiber discs (Whatman GF/A) and washed with 1% TCA. The rate of filtration is much slower with cold samples. The filters are placed flat on the bottom of the counting vials, dried for an hour at 80°C, and the protein is dissolved in 0.25 ml of Soluene 100 (Packard Instrument Company, Inc.) containing 5% water. When water is added to the Soluene, a white precipitate forms which dissolves in a few minutes with stirring. Dissolution of the protein is facilitated by elevated temperatures (e.g., capped vials can be incubated at 37°C overnight, or at 55°C for 1 hr). Only caps lined with polyethylene should be used. To each vial is added 15 ml of scintillation fluid containing 5 g of 2,5-diphenyloxazole (PPO) and 0.5 g of 1,4-bis-2-(5-phenyloxazolyl)benzene (POPOP) or 0.25 g dimethyl POPOP per liter of toluene. Both ^{14}C and ^{3}H are determined by liquid scintillation counting, and efficiencies of approximately 62 and 25% are obtained for ^{14}C and ^{3}H, respectively, with double-label counting (R. Jones and J. B. Lingrel, unpublished).

Samples containing hemoglobin are prepared for counting as described for globin, with the exception that the Soluene 100 used to dissolve the protein contains 0.1 vol of tert-butylhydroperoxide and no water. The peroxide decolorizes the hemoglobin and therefore eliminates the color quench (i.e., loss of counting efficiency resulting from absorption of β-particle-induced fluorescence by colored materials in the sample).

When a rabbit reticulocyte cell-free system is used for the assay of mouse hemoglobin mRNA, globin prepared from the combined cell-free incubations and mouse hemoglobin carrier is dissolved in 2-5 ml of 0.01 M $Na_2HPO_4 \cdot 7H_2O$ buffer (pH 6.9) containing urea and mercaptoethanol and applied to a column already equilibrated with the same buffer. The column is eluted with a 870-ml linear gradient from 0.01 M to 0.025 M $Na_2HPO_4 \cdot 7H_2O$ (pH 6.9). A 500-ml gradient also gives satisfactory results. A typical analysis of mouse mRNA, using a rabbit reticulocyte cell-free system, is shown in Figure 3 (R. Jones and J. B. Lingrel, unpublished). Mouse β chain elutes first and is labeled only when mouse mRNA is added to the incubation mixture. Mouse α chains have been shown to elute with rabbit chains, and their synthesis can not be determined by using the rabbit cell-free system [9]. The assay for mRNA activity is linear up to concentrations of 13 μg of mRNA per 0.5 ml of cell-free system.

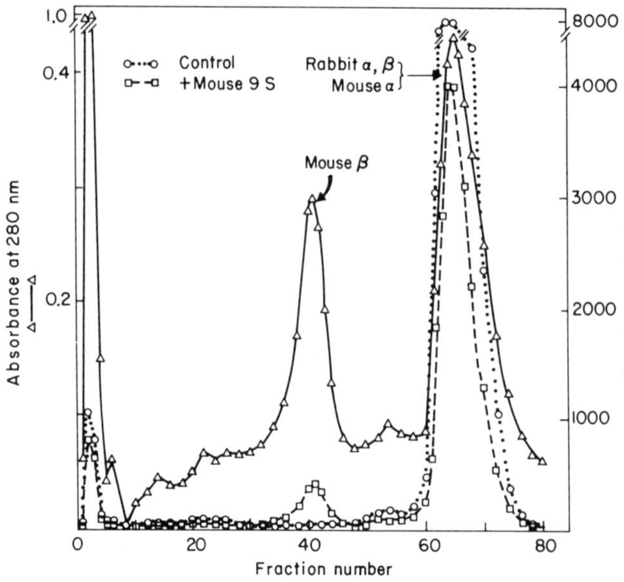

Fig. 3. Analysis of mouse hemoglobin mRNA using a rabbit reticulocyte cell-free system. Two 0.5-ml cell-free incubations were performed, one containing [^{14}C]L-leucine, 20 Ci/mole (1 μCi) and the other [^3H]L-leucine, 94 Ci/mole (4.7 μCi) and 7.5 μg of mouse mRNA. The two cell-free systems were incubated at 25°C for 2 hr. At the end of the incubation, the reaction mixtures were cooled to 0°C and pooled; 40 mg of mouse carrier hemoglobin were added, and globin was prepared. The globin was chromatographed using an 870-ml linear gradient from 0.01 to 0.025 M $Na_2HPO_4 \cdot 7H_2O$ (pH 6.9). The buffers contained 8 M urea and 50 mM β-mercaptoethanol. Five-milliliter fractions were collected.

Both mouse α- and β-chain mRNA activity can be determined by adding mouse mRNA to either a guinea pig or a duck cell-free system. With a mouse-guinea pig cross (Figure 4), mouse β chains elute after a small peak of labeled guinea pig protein and before guinea pig globin chains, while mouse α chains elute last and are completely resolved from the other chains (R. Jones and J. B. Lingrel, unpublished).

Figure 5 shows the analysis of mouse mRNA in a duck cell-free system. Duck globin elutes as two small peaks and one large one. The two small peaks are probably globin chains from the minor hemoglobin component, while the large peak is undoubtedly globin of the major duck hemoglobin. Although the identity of these peaks is unknown, they are present in duck globin and do not interfere with the separation of mouse α and β chains. The mouse β chain is completely resolved from duck globin, while the mouse α chain elutes close to the second duck peak (R. E. Lockard and J. B. Lingrel, unpublished).

Rabbit mRNA is assayed by adding it to a guinea pig cell-free system. The chains are eluted from the column by a combination of sodium ion concentration and pH gradient. As seen in Figure 6, guinea pig globin chains elute first, followed by rabbit chains which elute together. The combined rabbit α- and β-chain mRNA activity is determined in this assay (R. Jones and J. B. Lingrel, unpublished).

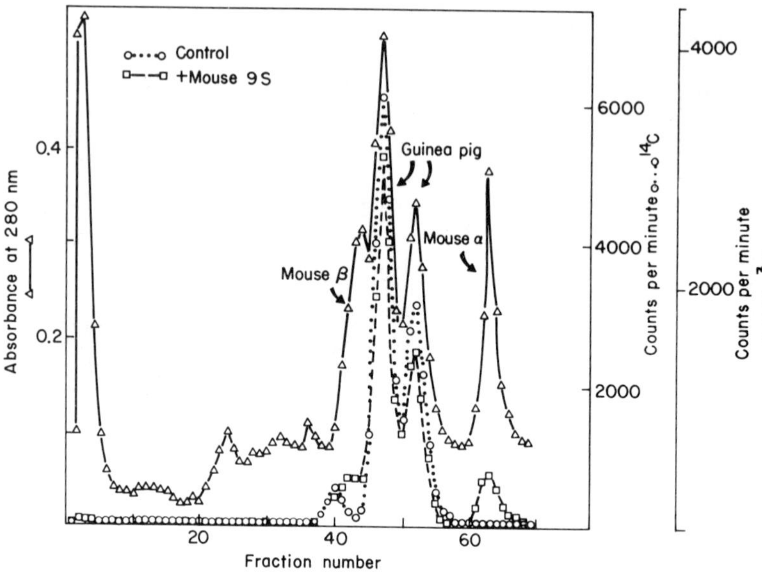

Fig. 4. Analysis of mouse hemoglobin mRNA using a guinea pig reticulocyte cell-free system. Two 0.5-ml cell-free incubations were performed. One contained [^{14}C]L-leucine, 20 Ci/mole (2 μCi) and the other [^{3}H]L-leucine, 94 Ci/mole (4.7 μCi) and 5 μg of mouse mRNA. The two cell-free systems were incubated at 25°C for 2 hr; at the end of the incubation, the reaction mixtures were cooled to 0°C, pooled, and 40 mg of mouse carrier hemoglobin were added. The globin was chromatographed using a linear 600-ml gradient from 0.01 to 0.025 M $Na_2HPO_4 \cdot 7H_2O$ (pH 6.8). The buffers contained 8 M urea and 50 mM β-mercaptoethanol. Five-milliliter fractions were collected.

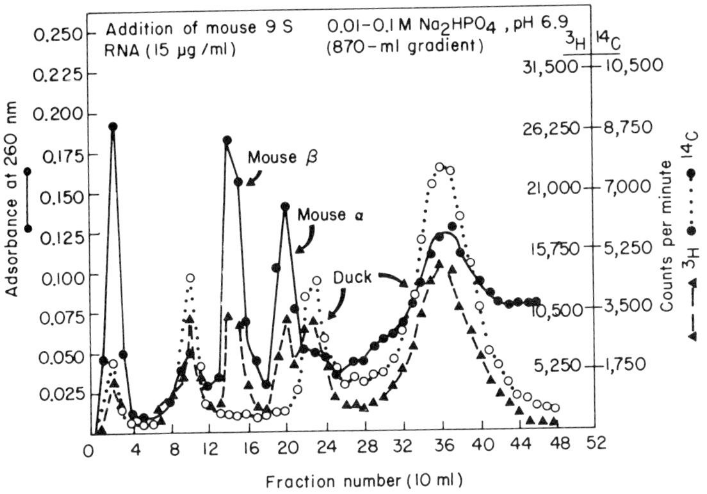

Fig. 5. Analysis of mouse hemoglobin mRNA using a duck reticulocyte cell-free system. Two 1-ml cell-free incubations were performed, one containing [^{14}C]L-leucine, 85 Ci/mole (8.5 μCi) and the other [^{3}H]L-leucine, 400 Ci/mole (40 μCi) and 15 μg of mouse hemoglobin mRNA. The two cell-free systems were incubated at 28°C for 1 hr, cooled to 0°C, combined, and globin was prepared after the addition of 90 mg of mouse carrier hemoglobin. One-half of the sample was chromatographed using an 870-ml linear gradient from 0.01 to 0.1 M $Na_2HPO_4 \cdot 7H_2O$ (pH 6.9). The buffers contained 8 M urea and 50 mM β-mercaptoethanol. Ten-milliliter fractions were collected.

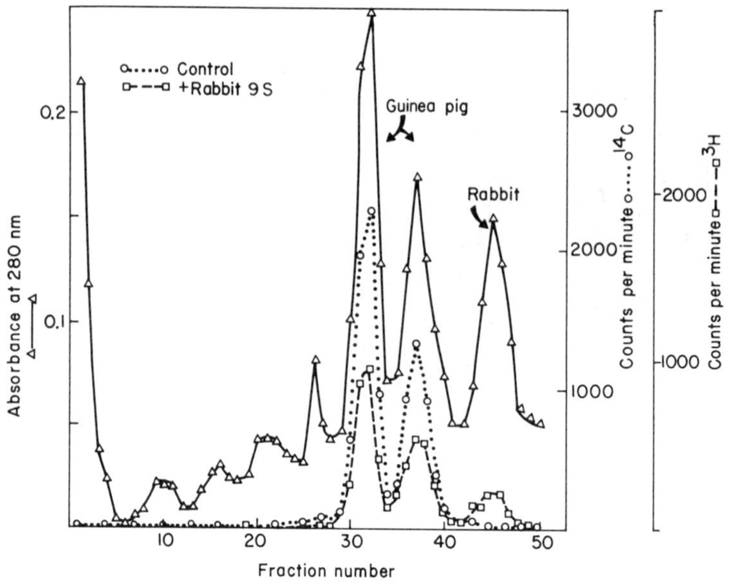

Fig. 6. Analysis of rabbit hemoglobin mRNA using the guinea pig reticulocyte cell-free system. Two 0.5-ml cell-free incubations were performed, one containing [^{14}C]L-leucine, 20 Ci/mole (1 μCi) and the other [^{3}H]L-leucine, 94 Ci/mole (4.7 μCi) and 10 μg of rabbit mRNA. After incubation for 2 hr at 25°C, the reaction mixtures were cooled to 0°C, combined, and 40 mg of carrier rabbit hemoglobin were added. The globin was chromatographed using a 750-ml linear gradient made by mixing 375 ml of 0.01 M $Na_2HPO_4 \cdot 7H_2O$ (pH 7.0) with 375 ml of 0.02 M $Na_2HPO_4 \cdot 7H_2O$ (pH 7.5). The solutions contained 8 M urea and 50 mM β-mercaptoethanol. Fractions of 7.5 ml were collected from the CM-52 carboxymethyl cellulose column.

Other mRNA and cell-free system combinations, and methods for separating globin chains, are certainly possible for the assay of mRNA but as yet have not been explored.

ACKNOWLEDGMENTS

The work reported from our laboratory was supported by U.S.P.H.S. research grant GM-10999, National Science Foundation grant GB-7993, and American Cancer Society grant E-479A. The author is a Research Career Development Awardee of the National Institutes of Health.

The following collaborators participated in the work reported from this laboratory: Raymond E. Lockard, Richard Jones, and Henry Burr.

REFERENCES

[1] F. Jacob and J. Monod, J. Mol. Biol., 3, 318 (1961).

[2] G. Marbaix and A. Burny, Biochem. Biophys. Res. Commun., 16, 522 (1964).

[3] A. Burny and G. Marbaix, Biochim. Biophys. Acta, 103, 409 (1965).

[4] G. Huez, A. Burny, G. Marbaix, and B. Lebleu, Biochim. Biophys. Acta, 145, 629 (1967).

[5] H. Chantrenne, A. Burny, and G. Marbaix, Progress in Nucleic Acid Research and Molecular Biology, 7, 173 (1967).

[6] M. J. Evans and J. B. Lingrel, Biochemistry, 8, 829 (1969).

[7] M. J. Evans and J. B. Lingrel, Biochemistry, 8, 3000 (1969).

[8] D. G. Laycock and J. A. Hunt, Nature, 221, 1118 (1969).

[9] R. E. Lockard and J. B. Lingrel, Biochem. Biophys. Res. Commun., 37, 204 (1969).

[10] G. Schapira, J. C. Dreyfus, and N. Maleknia, Biochem. Biophys. Res. Commun., 32, 558 (1968).

[11] S. M. Heywood and M. Nwagwu, Biochemistry, 8, 3839 (1969).

[12] L. H. Kedes and P. R. Gross, Nature, 223, 1335 (1969).

[13] T. W. Borun, M. D. Scharff, and E. Robbins, Proc. Natl. Acad. Sci. U.S., 58, 1977 (1967).

[14] D. Gallwitz and G. C. Mueller, F.E.B.S. Letters, 6, 83 (1970).

[15] J. Kruh and H. Borsook, J. Biol. Chem., 220, 905 (1956).

[16] G. Schapira, P. Padieu, N. Maleknia, J. Kruh, J. C. Dreyfus, Y. Alexandre, M. Benrubi-Dumont, L. Reibel, A. M. Thireau, and L. Tichonicky, J. Mol. Biol., 20, 427 (1966).

[17] J. Bishop, G. Favelukes, R. Schweet, and E. Russell, Nature, 191, 1365 (1961).

[18] J. B. Lingrel and H. Borsook, Biochemistry, 2, 309 (1963).

[19] G. Huez, A. Burny, G. Marbaix, and E. Schram, European J. Biochem., 1, 179 (1967).

[20] J. Temmerman and B. Lebleu, Biochim. Biophys. Acta, 174, 544 (1969).

[21] F. Labrie, Nature, 221, 1217 (1969).

[22] W. G. Lanyon, J. Paul, and R. Williamson, F.E.B.S. Letters, 1, 279 (1968).

[23] S. D. Adamson, E. Herbert, and W. Godchaux, III, Arch. Biochem. Biophys., 125, 671 (1968).

[24] W. V. Zucker and H. M. Schulman, Proc. Natl. Acad. Sci. U.S., 59, 582 (1968).

[25] M. Rabinovitz, M. L. Freedman, J. M. Fisher, and C. R. Maxwell, Cold Spring Harbor Symp. Quant. Biol., 34, 567 (1969).

[26] H. Borsook, E. H. Fisher, and G. Keighley, J. Biol. Chem., 229, 1059 (1957).

[27] M. Rabinovitz and J. M. Fisher, Biochim. Biophys. Acta, 91, 313 (1964).

Chapter 9

PREPARATION AND ASSAY OF RETICULOCYTE INITIATION FACTORS

David A. Shafritz, Philip M. Prichard, and W. French Anderson

Section on Human Biochemistry
National Heart and Lung Institute
National Institutes of Health
Bethesda, Maryland

I. INTRODUCTION 266

II. RIBOSOMES AND CRUDE RIBOSOMAL "HIGH-SALT WASH" FRACTION. 267

A. Preparative Procedures 267

B. Assay Procedures 272

III. FACTOR M_1. 276

A. Preparation. 277

B. Assay Procedures 280

IV. FACTOR M_2. 282

A. Preparation. 282

B. Assay Procedures 282

V. FACTOR M_3. 285

A. Preparation. 286

B. Assay Procedures 288

VI. SELECTION OF ASSAY PROCEDURE 289

VII. SUMMARY. 290

REFERENCES 291

I. INTRODUCTION

One of the most useful sources of material for studies on the mechanism of mammalian protein synthesis has been the rabbit reticulocyte. A major advantage of the reticulocyte system is the fact that greater than 90% of the product synthesized is hemoglobin, a protein that has been well characterized and can be easily identified. In the cell-free system, it has been found that when polyribosomes are washed with 0.5 M KCl, hemoglobin synthesis is limited to the completion of nascent globin chains, that is, chains that had already been initiated and had remained attached to the ribosomes during the washing procedure. When the ribosomal wash (0.5 M KCl) fraction is included in the reaction mixture with washed ribosomes, the product is complete, uniformly labeled hemoglobin chains. Thus this ribosomal high-salt wash fraction appears to be important in the initiation of new polypeptide chains [1,2]. This fraction has also been shown to stimulate the formation of polysomes from washed reticulocyte monosomes and has been separated into two protein fractions by sucrose gradient centrifugation [3,4].

In our laboratory the crude reticulocyte, high-salt ribosomal wash fraction has been separated into at least three specific initiation factors, M_1, M_2, and M_3, M referring to their mammalian origin [5—7]. The basic biological properties of these factors and their initial preparation starting with the ribosomal wash fraction are similar to those of bacterial systems (see

Chapter 6, Vol. 1). Two of the mammalian factors, M_1 and M_2, can bind reticulocyte Met-tRNA (a species of methionyl-tRNA that can be formylated by Escherichia coli transformylase [8]) to reticulocyte ribosomes at low Mg^{2+} concentration, using as messenger the initiator codon triplet AUG [9]. The third factor, M_3, appears to be required specifically for translation of hemoglobin mRNA at low Mg^{2+} concentration [6]. A factor that may be similar to M_3 (EF_3, prepared from chicken reticulocyte or muscle ribosomes) has been reported to bind hemoglobin mRNA to washed chicken reticulocyte ribosomes [10]. In contrast to bacterial systems, the mammalian cell-free system with artificial messenger does not require formylation of the initiator tRNA substrate, although the initiation process can be accomplished with N-formylmethionyl-tRNA. This observation is not surprising in view of the apparent absence of transformylase enzyme activity in the cytoplasm of eukaryotic cells. In addition to studies of reticulocyte initiation factors, evidence that natural initiation in eukaryotic systems occurs with unformylated Met-$tRNA_F$ has been obtained by several research groups [11—26].

II. RIBOSOMES AND CRUDE RIBOSOMAL HIGH-SALT WASH FRACTION

A. Preparative Procedures

1. Preparation of Lysate.

Techniques for inducing reticulocytosis and obtaining blood with a high reticulocyte count are described in the literature

[27] (see Chapters 2 and 8). In our laboratory [28] reticulocytosis is induced in immature New Zealand white rabbits (1.5—2.0 kg) by six successive daily subcutaneous injections of phenylhydrazine [0.25 ml/kg body weight of a 2.5% (w/v) solution of phenylhydrazine hydrochloride, neutralized to pH 7.0 by NaOH]. For optimal results the daily dosage of phenylhydrazine is adjusted for change in body weight of the rapidly growing rabbits. On the first day the animals are also given (intramuscularly) 1 ml of vitamin B_{12}—folate solution [10 mg of cyanocobalamin, 100 mg of folic acid in 100 ml of 0.9% (w/v) NaCl, adjusted to pH 7.0 with NaOH]. The rabbits are rested on the seventh day, sedated on the eighth day with 1.0 ml (intraperitoneal) of pentabarbital (65 mg/ml), and bled by direct cardiac puncture using a 50-ml syringe containing 1 ml of heparin (1000 units/ml). The blood (50—60 ml per rabbit) is pooled into chilled, heparinized 250-ml centrifuge bottles and packed by centrifugation at 10,000 x g for 10 min. This and all subsequent procedures are performed at 2—4°C. The cells are washed twice by suspension in 2 vol of wash solution (0.14 M NaCl, 0.05 M KCl, 0.005 M $MgCl_2$) followed by centrifugation at 10,000 x g for 10 min. After the second wash step, the packed cells are lysed by suspension in 4 vol of lysing solution [2 mM $MgCl_2$, 1 mM dithiothreitol (DTT), 0.1 mM EDTA, neutralized to pH 7.0 by NaOH]. The suspension is swirled rapidly for 2—3 min and then centrifuged at 10,000 x g for

20 min. The supernatant fraction (lysate) is removed from the pelleted cell debris and can be used immediately or stored in liquid nitrogen for several months without loss of activity.

2. Preparation of Unwashed Polysomes and Crude Supernatant.

Crude unwashed reticulocyte polysomes are obtained from lysate by centrifugation at 105,000 X g for 2 hr. The upper three-fourths of the supernatant fraction (containing predominantly elongation factors, synthetases, tRNA, and hemoglobin) is carefully removed and recentrifuged at 175,000 X g for 2.5—3 hr to remove any residual contamination from ribosomal subunits.

3. Preparation of Washed Ribosomes and Crude Ribosomal Wash Fraction

The surface of the crude polysome pellet is washed three times with standard sucrose solution A (0.25 M sucrose, 1 mM DTT, 0.1 mM EDTA, neutralized to pH 7.0 with KOH) and the pellets are gently suspended to a final concentration of approximately 125 A_{260} units/ml. When preparing well-washed ribosomes (for the best enzyme dependencies), a starting concentration of unwashed polysomes of 125 A_{260} units/ml is used; when preparing a highly concentrated crude ribosomal high-salt wash fraction, which also produces ribosomes with maximal natural endogenous messenger activity, a starting concentration of unwashed ribosomes of 250 A_{260} units/ml is used. The most rapid and effective means of gently solubilizing the polysomal pellets is to scrape them from the side wall of the centrifuge tube with a glass rod, pool them in a beaker, and mix them at the slowest possible speed with a small magnetic stirring bar. Suspension can also be accomplished by twirling the pooled ribosomal pellets in a conical centrifuge tube placed in a bucket of ice. Both of these procedures are used to minimize the shearing forces placed on the very long and thin strands of hemoglobin mRNA. After the polysomes are suspended 4.0 M KCl is added dropwise over several minutes, with stirring, to a final concentration of 0.5 M in KCl (0.1 ml of 4.0 M KCl per 0.7 ml of polysome suspension), and the solution is stirred gently for several hours for well-washed ribosomes, or 10 min for ribosomes with maximal endogenous messenger activity. Additional Mg^{2+} ion should not be added during this step, since the ribosomal wash fraction obtained already contains an apparent divalent cation concentration in excess of 10 mM (discussed below). In fact, the ability to obtain ribosomes dependent on the high-salt ribosomal wash fraction appears to be adversely affected by the addition of Mg^{2+} during the wash procedure.

The 0.5 M KCl-washed polysomes, reported to be predominantly in the form of ribosomal subunits [29], are clarified by centrifugation at 10,000 $\times$ g for 10 min, and the ribosomes are pelleted by centrifugation at 258,000 $\times$ g for 2—3 hr. The upper three-fourths of the soluble ribosomal wash (0.5 M KCl) fraction is carefully removed and stored in liquid nitrogen or used directly for the preparation of M factors. In addition to initiation factors, this fraction also contains elongation factors, aminoacyl-tRNA synthetases, tRNA, and small quantities of intact hemoglobin mRNA.

The surface of the ribosomal pellets and the side walls of the centrifuge tubes are carefully rinsed three times with standard sucrose solution B [0.25 M sucrose, 20 mM tris—HCl (pH 7.5), 40 mM KCl, 5 mM MgCl, 1 mM DTT, 0.1 mM EDTA]. The pellets are carefully drained and then gently suspended to approximately 200 A_{260} units/ml in standard sucrose solution B; 4 M KCl is then added to a final concentration of 0.5 M and the solution is stirred gently for 1 hr. These twice high-salt-washed ribosomes are pelleted at 258,000 X g for 2—3 hr and then suspended in standard sucrose solution B to a final concentration of approximately 300—500 A_{260} units/ml. The ribosomal suspension is clarified by centrifugation at 10,000 X g for 10 min and has a final A_{280}/A_{260} ratio of approximately 0.54. Although the second high-salt wash step results in an increased ribosomal dependence on added enzyme factors and message, it is not usually necessary when obtaining preliminary results and in addition, the second wash step causes a 50—60% decrease in total ribosomal activity. It is important to note that the most active preparations of twice high-salt-washed ribosomes are obtained when all the steps following the preparation of lysate are performed as a single unit. The ribosomes are stored in small aliquots in liquid nitrogen without significant loss of activity over 6 months.

B. Assay Procedures

The crude ribosomal wash fraction can be assayed by its ability to stimulate cell-free, endogenous messenger-dependent hemoglobin synthesis with high-salt-washed ribosomes at low Mg^{2+} concentration [1,2], or by its ability to lower the Mg^{2+} concentration optimum for cell-free, poly U-dependent polypeptide synthesis (the "Mg^{2+} shift") [1,5,7].

1. Cell-Free Hemoglobin Synthesis at Low Mg^{2+} Concentration

This assay, under appropriate conditions, gives a 30-fold stimulation by the ribosomal wash fraction. The most active components are ribosomes and ribosomal wash prepared at a polysome concentration of 250 A_{260} units/ml. Total protein synthesis is routinely measured by the incorporation of radioactive amino acid into material insoluble in hot 10% trichloroacetic acid (TCA). The protein product appears to be uniformly labeled after approximately 7 min of incubation in the presence of the ribosomal wash fraction. In this system protein synthesis is linear for 45—50 min, and it has been estimated from turnover studies that each active hemoglobin messenger is probably utilized a minimum of 12 times during the course of the incubation. Since the crude ribosomal wash fraction contains near-saturating quantities of "supernatant" proteins, 0.5 mg or less of crude supernatant must be added to each assay. If the ribosomal wash fraction is dialyzed prior to use (to remove residual amino acids and other small molccules), the Mg^{2+} concentration optimum for endogenous messenger-dependent protein synthesis is 4.5 mM (i.e., added Mg^{2+}) rather than 3 mM.

Incubations, in a total volume of 50 μl, are performed at 37°C for 20—30 min and contain 20 mM tris—HCl (pH 7.5), 80 mM KCl, 3.0 mM $MgCl_2$, 0.2 mM GTP (neutralized to pH 7.0 with KOH), 1 mM ATP (neutralized to pH 7.0 with KOH), 3 mM phosphoenolpyruvate (neutralized to pH 7.0 with KOH), 0.3 IU pyruvate kinase (rabbit muscle), 0.08 mM (each) of 19 [^{12}C]amino acids, 0.08 mM [^{14}C]valine (or [^{3}H]leucine, and so on), 1 mM DTT, 0.05 A_{260} unit of unfractionated reticulocyte tRNA, 0.2 A_{260} unit of washed reticulocyte ribosomes (rate limiting for the reaction), 0.5 mg of supernatant protein, and appropriate amounts of crude ribosomal wash protein (approximately 100 μg of protein to saturate the assay). The reaction is stopped by the addition of 2 ml of ice-cold 10% TCA. The reaction mixtures are then heated to 90—95°C for 15 min to deacylate the residual tRNA, cooled for 10 min in ice, and the precipitated protein is collected on nitrocellulose filters (Millipore Corporation, type HA, 0.45-μ pore size, 25-mm diameter). Glass-fiber filters can also be used for this assay. The filters are washed with 20 ml of ice-cold 5% TCA, dried under an infrared lamp, and counted by liquid scintillation in 10 ml of a Liquifluor—toluene counting solution.

2. The Poly-U Mg^{+} Shift

This assay, which gives a 5- to 6-fold stimulation by high-salt wash protein, can be performed either with [^{14}C]phenylalanine [1,5], crude aminoacyl-tRNA synthetases, deacylated tRNA, and an ATP-generating system, or with previously acylated [^{14}C]Phe-tRNA [5,7]. In the former case the reaction conditions are identical to those described for hemoglobin synthesis except that 0.5—0.7 A_{260} unit of poly U are added to the reaction mixture. The reaction is linear for 45—60 min under the conditions used; protein synthesis is optimal at 5.5 mM Mg^{2+} in the presence of ribosomal wash, and 10 mM Mg^{2+}, in the absence of ribosomal wash. The reaction is almost completely dependent on poly U at these Mg^{2+} concentrations but is not dependent on poly U at 3 mM Mg^{2+}, where endogenous messenger for hemoglobin is translated.

With [^{14}C]Phe-tRNA as substrate, the assay is essentially the same as that originally reported by Lucas-Lenard and Lipmann for *E. coli* initiation factors F_1 and F_2 [34], except that in the reticulocyte system N-acetyl-Phe-tRNA is not required. (Methods can be found in the literature for preparing mammalian tRNA [2,28,31], chromatographic separation of tRNA species [9,32], acylation [2,28,31], formylation [8], and N-acetylation [33].) The reaction is very rapid, and protein synthesis is measured in a 2- to 4-min incubation or until the [^{14}C]Phe-tRNA substrate becomes rate limiting. Once acylated tRNA becomes rate limiting, the effect of M factors on polypeptide synthesis is markedly diminished. Incubations, in a total volume of 50 µl, are performed at 37°C for 2 min and contain 20 mM tris—HCl (pH 7.5), 100 mM KCl, 1 mM GTP, 3 mM phosphoenolpyruvate, 0.3 IU pyruvate kinase, 1 mM DTT or 20 mM β-merceptoethanol, 0.2 A_{260} unit of washed ribosomes, 0.5 mg of supernatant protein, 0.5 A_{260} unit of poly U, 7—10 pM [^{14}C]Phe-tRNA (2% charged; prepared from either *E. coli* or unfractionated reticulocyte tRNA), saturating quantities of ribosomal wash protein, and the appropriate Mg^{2+} concentration. This reaction can be performed either in the presence or absence of phosphoenolpyruvate and pyruvate kinase. In the absence of these substances, which along with ATP can bind Mg^{2+}, total activity is reduced by approximately 35%. The Mg^{2+} shift caused by the addition of the ribosomal wash fraction is from 8 to 3 mM Mg^{2+}. In the presence of 3 mM phosphoenolpyruvate (and pyruvate kinase), the Mg^{2+} shift caused by the addition of ribosomal wash is from 10 to 5 mM Mg^{2+} (Figure 1). The Mg^{2+} shift is also more distinct when *E. coli*, rather than rabbit reticulocyte, is the source of [^{14}C]Phe-tRNA. In all assays for poly U-dependent polypeptide synthesis, the most active ribosome preparations are obtained from polysomes washed once with 0.5 M KCl (or 1 M KCl) at a concentration of 100—125 A_{260} units/ml.

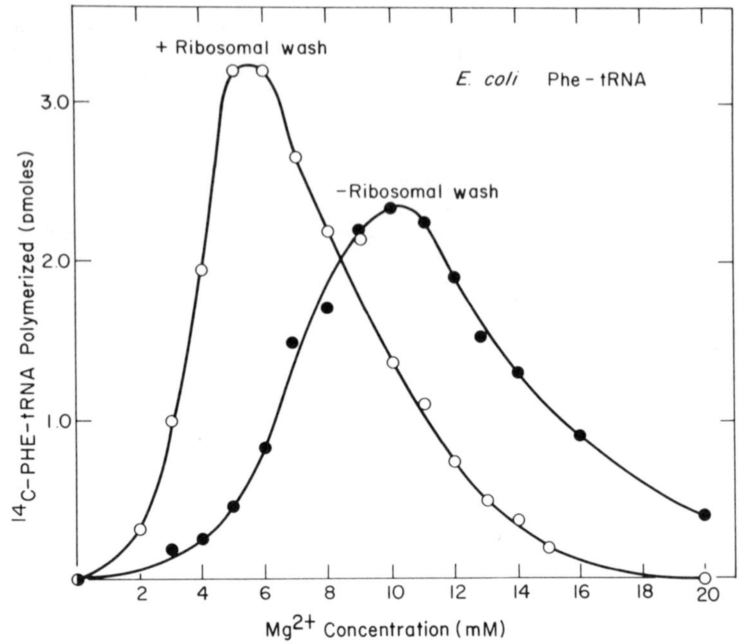

Fig. 1. The effect of the (0.5 M KCl) ribosomal wash fraction on the Mg^{2+} concentration optimum for poly U-directed polypeptide synthesis. See Ref. 5 for experimental details.

III. FACTOR M_1

M_1 is a protein which, under appropriate conditions, can enzymically bind either free or N-blocked aminoacyl-tRNA to washed reticulocyte ribosomes at low Mg^{2+} concentration. This protein is the most stable of the reticulocyte factors and has been completely separated from the other M factors, as well as reticulocyte T_1 (the aminoacyl-tRNA binding protein), by a combination of DEAE-cellulose and Sephadex G-150 chromatography [7].

A. Preparation

A 150-ml portion of ribosomal wash protein (4—5 mg/ml) is precipitated by the stepwise addition of solid, enzyme grade $(NH_4)_2SO_4$ (Schwarz/Mann) with constant stirring at 0°C. The pH usually does not drop below 6.5—6.8 but can be adjusted to 7.0—7.4 by the addition of ammonium carbonate, 1 g/40 g of $(NH_4)_2SO_4$. The protein fraction precipitating between 30 and 65% $(NH_4)_2SO_4$ saturation is collected and washed once with 68% $(NH_4)_2SO_4$ in buffer B [10 mM tris—HCl (pH 7.5), 50 mM KCl, 1 mM $MgCl_2$, 1 mM DTT, 0.1 mM EDTA]. A fractional $(NH_4)_2SO_4$ precipitation can also be performed to partially separate M_1 and M_2, although this procedure has not been generally used. The majority of M_2 precipitates between 35 and 50% $(NH_4)_2SO_4$ saturation, and most of the M_1 between 45 and 65% $(NH_4)_2SO_4$ saturation. The $(NH_4)_2SO_4$ fraction is dialyzed for 6 hr against two changes of buffer B (2 liters for each change) and centrifuged to remove particulate matter. This material can be stored in liquid nitrogen or used directly for the preparation of M_1 and M_2.

Microgranular DEAE-cellulose (Whatman DE-52) or fibrous DEAE-cellulose (Whatman DE-23) can be used to separate M_1 and M_2. Fine particles are removed from the cellulose by flotation and decantation, followed by cycling through 0.5 M NaOH and 0.5 M HCl according to standard procedures [35]. The DEAE-cellulose is equilibrated with buffer B, poured into a 1.5 X

25 cm column, and further equilibrated for 48 hr before use. Dialyzed crude M factors at a protein concentration of 10—20 mg/ml are placed over this column, and 5-ml fractions are collected. The column is washed with 3—5 vol (75—125 ml) of starting buffer (buffer B, 0.05 M KCl), and a 500 ml (total) linear gradient from 0.05 M KCl (250 ml) to 0.4 M KCl (250 ml) is started; 5-ml fractions are collected. For best results, a Varigrad gradient maker, which gives perfectly linear gradients, is used. M_1 usually elutes directly from the column before the KCl gradient is started (slightly variable with different batches of DEAE-cellulose); the major peak of M_2 activity (M_{2_A}) elutes at approximately 0.26 M KCl, as shown in Figure 2.

M_1 can be further purified 5 fold and completely separated from T_1 by Sephadex G-150 chromatography. The size of the Sephadex column is selected on the basis of the amount and volume of M_1 available. For a 2.5 X 90 cm column of Sephadex G-150, 3—4 ml of M_1 at concentrations up to 50 mg protein per milliliter can be used. The column is equilibrated with buffer A [10 mM tris—HCl (pH 7.5), 100 mM KCl, 1 mM $MgCl_2$ 1 mM DTT, 0.1 mM EDTA]. The flow rate is maintained at 30—40 ml/hr, and 3-ml fractions are collected. M_1 elutes from this column just prior to the hemoglobin band. M_1 can also be purified by carboxymethyl cellulose chromatography at pH 6.5, but results have been generally poor and highly variable in terms of the amount of M_1 recovered.

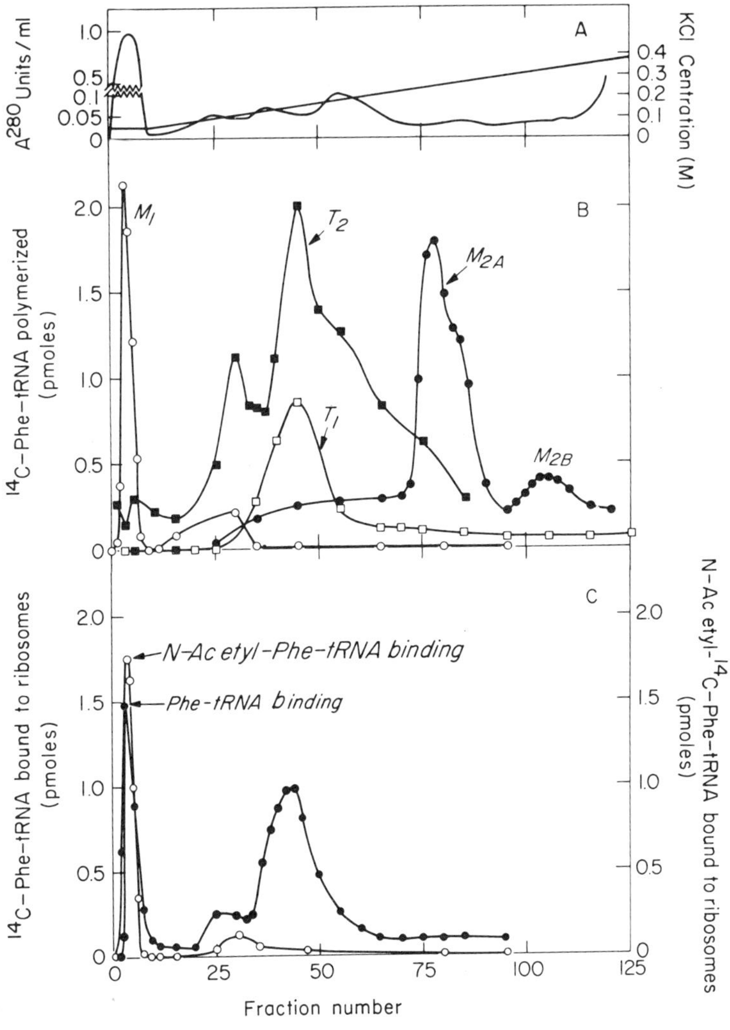

Fig. 2. Fractionation of M_1 and M_2 by DEAE-cellulose chromatography. See Ref. 7 for experimental details.

B. Assay Procedures

M_1 can be assayed by the poly-U Mg^{2+} shift or by its requirement for hemoglobin synthesis dependent on natural messenger at low Mg^{2+} concentration (see above) [5—7]. A more rapid and direct assay for M_1 is its ability to bind both [^{14}C]Phe-tRNA and N-acetyl-[^{14}C]Phe-tRNA to washed reticulocyte ribosomes, at low Mg^{2+} concentration, using poly U message [7].

This assay, which is a modification of the aminoacyl-tRNA binding assay originally reported by Nirenberg and Leder [36], can also be performed with [^{3}H]fMet-tRNA or with [^{3}H]Met-tRNA (in this case also requiring M_2) using the triplet AUG, or the randomly ordered polynucleotide poly AUG as message (see Figure 3).

Incubations, in a total of 50 μl, are performed at 23°C for 2—5 min and contain 20 mM tris—HCl (pH 7.5), 100 mM KCl, 5 mM Mg^{2+}, 0.2—0.5 mM GTP, 1 mM DTT, 1.0 A_{260} unit of washed ribosomes, 6—8 pM [^{14}C]Phe-tRNA (or [^{3}H]fMet-tRNA), 0.05 A unit of poly U (or 0.15 A_{260} unit of AUG or 0.4 A_{260} unit of poly AUG) and the appropriate amount of M_1. The reaction is stopped by the addition of 3 ml of ice-cold buffer solution [20 mM tris—HCl (pH 7.5), 100 mM KCl, 5 mM $MgCl_2$]. The reaction mixture is immediately collected on a nitrocellulose filter; the filter is washed three times with 3 ml of cold buffer, and then counted by liquid scintillation.

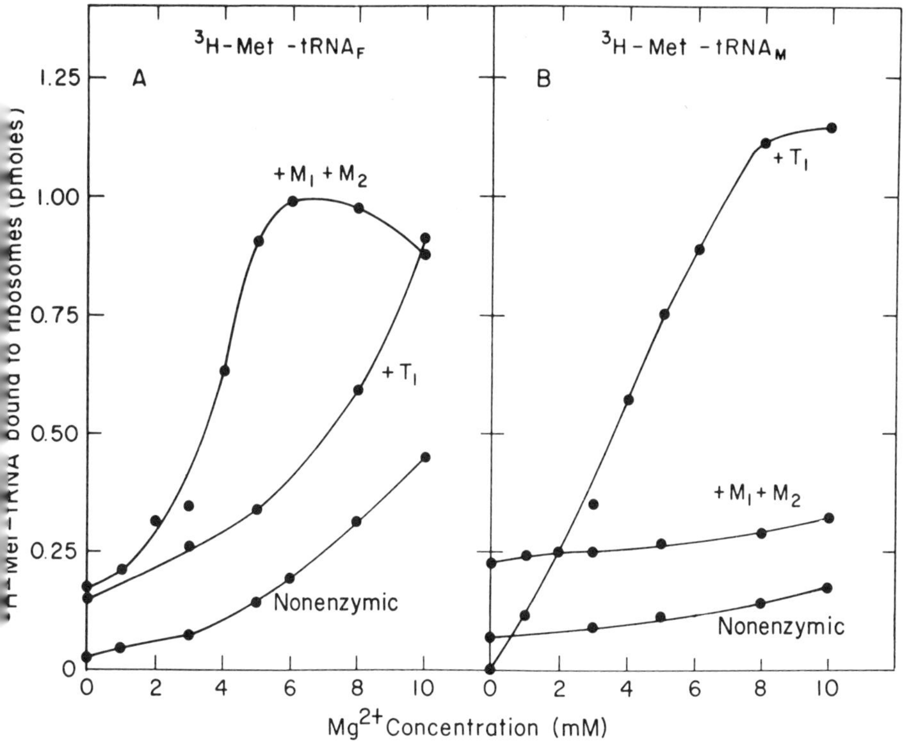

Fig. 3. Binding of [^{3}H]Met-tRNA$_F$ and [^{3}H]Met-tRNA$_M$ to reticulocyte ribosomes in response to M factors and T_1. See Ref. 9 for experimental details.

By the use of this assay, M_1 can be distinguished from T_1, since T_1 binds only [^{14}C]Phe-tRNA and [^{3}H]Met-tRNA , whereas M_1 binds [^{14}C]Phe-tRNA, N-acetyl-[^{14}C]Phe-tRNA, [^{3}H]Met-tRNA$_F$ (in the presence of M_2), [^{3}H]Met-tRNA$_M$ (to a small degree), and [^{3}H]fMet-tRNA.

IV. FACTOR M_2

A. Preparation

M_2 is prepared by the same procedures as M_1. M_2 (M_{2A}) elutes from a DEAE-cellulose column at approximately 0.26 M KCl. A second, minor peak of M_2 activity (M_{2B}) elutes at about 0.32 M KCl (see Figure 2); however, this elution pattern is variable. At present, it is not certain whether these separate peaks of M_2 activity represent two separate factors or two nonidentical subunits of a common enzyme function. From preliminary experiments with Sephadex G-200 chromatography, M_{2A} appears to be much larger in molecular weight than M_{2B}; M_{2A} emerges in the void volume and M_2 B is eluted behind M_1.

B. Assay Procedures

M_2 can be assayed by the poly-U Mg^{2+} shift or hemoglobin synthesis dependent on endogenous messenger. The tRNA-binding assay with [^{3}H]Met-$tRNA_F$ (or unfractionated [^{3}H]Met-tRNA) can also be used, since it requires M_2.

Another assay for M_2 activity (which can also be used for M_1), is the ability of this factor to lower the Mg^{2+} concentration optimum for puromycin peptide release with either [^{14}C]Phe-tRNA or [^{3}H]Met-tRNA, the "puromycin Mg^{2+} shift" [37]. This assay is a modification of the puromycin release assay with initiator tRNA reported for _E. coli_ by Leder and Bursztyn [38]. As shown in Figure 4, in the absence of M factors puromycin peptide release is optimal at 6 mM Mg^{2+} (for [^{14}C]Phe-tRNA or [^{3}H]Met-tRNA$_M$), while in the presence of M_1 plus M_2 the Mg^{2+} optimum is 3 mM (for [^{14}C]Phe-tRNA, N-acetyl-[^{14}C]Phe-tRNA, [^{3}H]Met-tRNA$_F$, or [^{3}H]fMet-tRNA).

Incubations, in a total volume of 50 or 100 μl, are performed at 30°C for 10 min and contain 20 mM tris--HCl (pH 7.5), 100 mM KCl (for T factors) or 150 mM KCl (for T plus M factors), 3 mM Mg^{2+}, 0.5 mM GTP, 1 mM DTT, 0.5 mM puromycin (neutralized to pH 7.0), 1.0 A_{260} unit of washed ribosomes (twice-washed ribosomes give better dependencies), 10—15 pM [^{14}C]Phe-tRNA (or N-acetyl-[^{14}C]Phe-tRNA or [^{3}H]Met-tRNA$_F$ or [^{3}H]fMet-tRNA), 0.5 A_{260} unit of poly U (or 0.15 A_{260} unit of AUG), M_1 and M_2 (or M_{2A} plus M_{2B}). For all substrates the reaction is stimulated by T_1, and for [^{14}C]Phe-tRNA and [^{3}H]Met-tRNA$_M$ (at 6 mM Mg^{2+}), the reaction is stimulated by T_2. The reaction is stopped by the addition of 0.90—0.95 ml of 0.1 M potassium phosphate buffer (pH 8.0) and puromycin peptides are extracted into 3 ml of ethyl acetate. After the ethyl acetate is added, the reaction mixture is mixed briskly for 30 sec with a vortex mixer and then centrifuged for 2 min in a clinical desk-top centrifuge. Two milliliters of the ethyl acetate fraction [removed from the upper layer, carefully avoiding the lower (aqueous) layer] are counted by liquid scintillation in 10 ml of Bray's solution [39]. Since not all of the puromycin peptide is extracted into the ethyl acetate layer, the appropriate

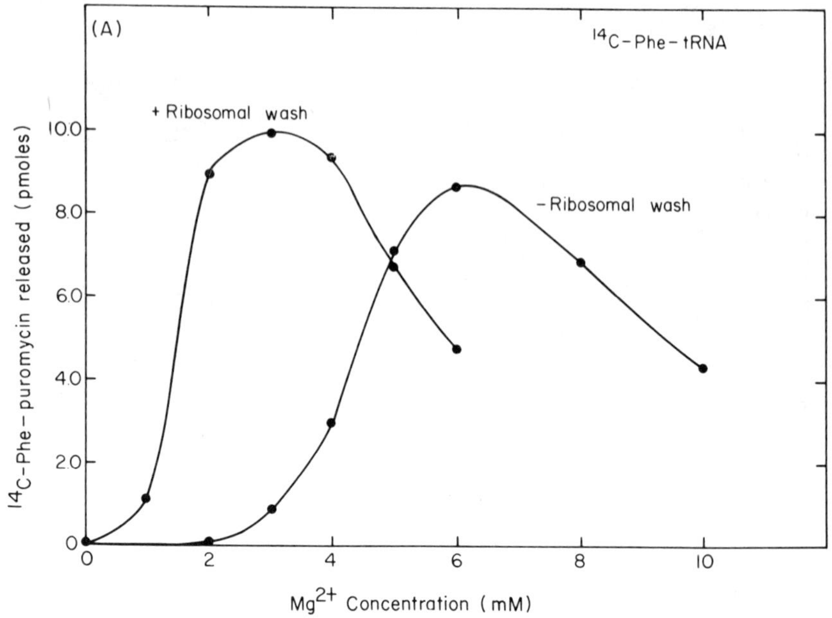

Fig. 4A. [^{14}C]Phe-puromycin release assay using reticulocyte ribosomes in the presence and absence of the ribosomal wash fraction. See Ref. 37 for experimental details.

---

correction should be made when calculating picomoles of product [31]. During the course of the incubation, there is no evidence for either formylation or deformylation of the substrate, so that when formylated (or acetylated) substrates are used the ethyl acetate extraction can be performed at pH 5.5. When [^{3}H]Met-tRNA$_M$ is used as substrate, poly AUG (0.4 A_{260} unit) must be used in place of the triplet AUG, since the triplet is highly inefficient with Met-tRNA$_M$.

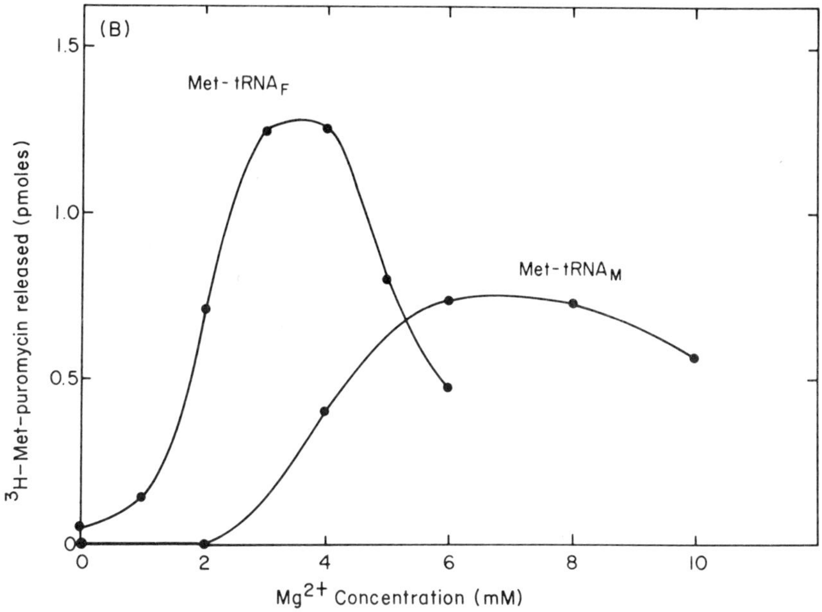

Fig. 4B. [^{3}H]Met-puromycin release assay using reticulocyte ribosomes with either Met-tRNA$_F$ (with M plus T factors) or Met-tRNA$_M$ (with T factors only). See Ref. 37 for experimental details.

V. FACTOR M_3

M_3 has been isolated from the ribosomal wash fraction by a combination of several procedures [6]. Once separated from the M_1 and M_2 factors, it appears to be less stable; attempts to stabilize and further purify M_3 have been only partially successful. Adsorption of the crude ribosomal wash from rabbit reticulocytes to DEAE-cellulose at low KCl concentration reduces M_3 activity. A DEAE-cellulose batch method has been developed, however, that selectively adsorbs M_2, leaving M_3 plus M_1 in the soluble fraction. M_1 elutes from DEAE-cellulose below 0.1 M KCl, whereas M_2 elutes from DEAE-cellulose at 0.26 M KCl (shown in Figure 1). M_3 (of very low activity) elutes at approximately 0.18 M KCl, so that 0.2 M KCl can be used for the batch procedure.

A. Preparation

DEAE-cellulose is equilibrated with 0.2 M KCl, 20 mM tris—HCl (pH 7.5), 1 mM DTT, and 0.1 mM EDTA (neutralized to pH 7.0 with NaOH). One milliliter of the ribosomal wash fraction (10 mg/ml protein in 0.5 M KCl) is diluted 2.5 times with the above buffer to give a final KCl concentration of 0.2 M and a protein concentration of approximately 4 mg/ml. This diluted wash fraction is then added to DEAE-cellulose that has been equilibrated with buffer and pelleted in a 12-ml conical centrifuge tube. The amount of DEAE-cellulose required for this step is variable with different preparations of DEAE-cellulose and ribosomal wash. From 0.5 to 2.5 ml of packed DEAE-cellulose have been used effectively to produce an active M_3 plus M_1 fraction that is up to 10-fold dependent on added M_2 for hemoglobin synthesis. The solution is thoroughly mixed with the DEAE-cellulose and is placed on ice for 10 min with occasional mixing. The DEAE-cellulose is sedimented by centrifugation at 2000 X g for 5 min. The pellet contains bound M_2, and the supernatant contains M_3 plus M_1. The supernatant fraction is collected and centrifuged again to remove small DEAE-cellulose particles. The M_3 contaminated with M_1, as prepared in this manner, is reasonably stable and can be stored in liquid nitrogen.

Sephadex G-200 chromatography can be used to separate M_3 plus M_2 from M_1, as shown in Figure 5. M_3 contaminated with M_2, as prepared in this manner, is less stable and has a half-life of approximately 6 hr at 0 or 25°C. However, concentration by ultrafiltration and storage in liquid nitrogen appears to allow retention of the amount of activity present at the time of freezing.

One milliliter of the ribosomal wash fraction is applied to a 0.9 X 50 cm Sephadex G-200 column that has been equilibrated with 0.1 M KCl, 3 mM tris—HCl (pH 7.5), 1 mM DTT, and 0.1 mM EDTA. The column is run at a slow flow rate (about 12 ml/hr) and can be eluted either at room temperature or at 4°C. Chromatography at 2—4°C appears to improve factor stability. M_3 and M_2 elute from the column at the void volume, whereas M_1 is included and elutes just prior to the hemoglobin peak (Figure 5).

A combination of the DEAE-cellulose batch method and Sephadex G-200 chromatography is used to separate M_3 from both M_1 and M_2. M_3 plus M_1 is prepared by the DEAE-cellulose batch method and concentrated by vacuum dialysis at 4°C (collodion bag filter apparatus). This material is then chromatographed

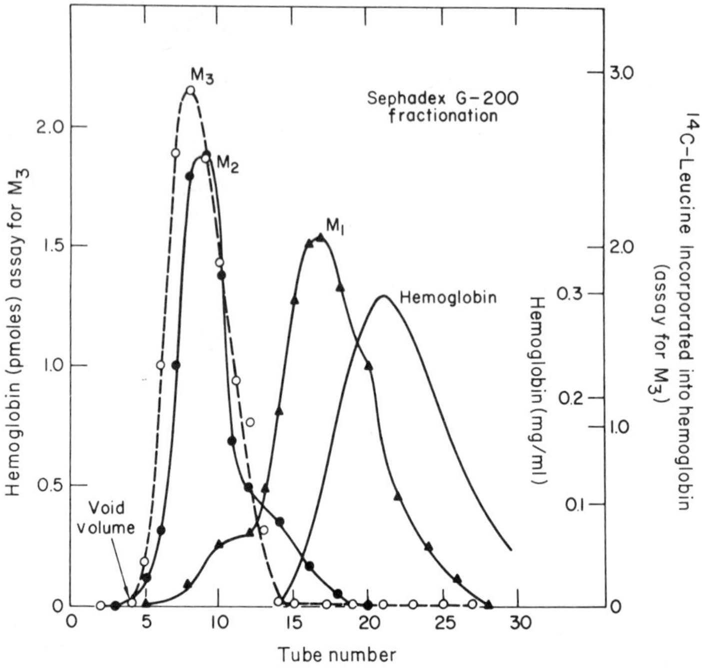

Fig. 5. Fractionation of the M factors by Sephadex G-200 column chromatography. See Ref. 6 for experimental details.

---

on Sephadex G-200 in the same manner as described above for the fractionation of ribosomal wash protein. M_3 elutes at the void volume and is well separated from M_1. Protein synthesis at 4.5 mM Mg^{2+} with this preparation of M_3 is stimulated up to 10-fold by the addition of M_1 and M_2.

B. Assay Procedures

M_3 is assayed, in the presence of M_1 and M_2, in the natural messenger-dependent system, whereas M_1 and M_2 can be assayed independently of M_3 for the poly U-directed synthesis of polyphenylalanine. The conditions for the assay of M_3 are the same as for the ribosomal (0.5 M KCl) wash fraction.

The Mg^{2+} optimum in the system dependent on natural messenger has been shown to be approximately 3 mM of added Mg^{2+}. Dialysis or Sephadex chromatography of the wash fraction may be used to separate the Mg^{2+} from the enzyme fraction. When M factors or dialyzed ribosomal wash are used, the Mg^{2+} concentration optimum changes to 4.5 mM. Since this requirement is very precise, it is critical that the correct Mg^{2+} concentration be used in each assay.

VI. SELECTION OF ASSAY PROCEDURE

For the initial isolation of an active high-salt ribosomal wash fraction from rabbit reticulocytes, the most sensitive assay is the stimulation of protein synthesis dependent on endogenous messenger (hemoglobin) at low Mg^{2+} concentration. The marked dependence on M factors under these conditions appears to indicate that protein synthesis is greatly stimulated by the ability of these factors to initiate new polypeptide chains. However, when we isolate specific initiation factors from the ribosomal wash fraction, the poly-U Mg^{2+} shift assay has been quite useful. Although this assay may not be a true model for mammalian initiation, it is dependent only on factors M_1 and M_2 and has the advantages of being more rapid (with acylated [^{14}C]Phe-tRNA as substrate) and less complex than the natural messenger-dependent assay since it requires one less factor (i.e., there is no requirement for M_3).

Once initiation factors have been isolated and partially purified using the very sensitive protein synthesis assays, it is then not difficult to demonstrate their function in aminoacyl-tRNA binding and peptide bond formation with Met-tRNA (or Phe-tRNA). In each instance we have found that the poly U-dependent system is more active than AUG- or natural messenger-dependent activity, although the latter appears to be more appropriate for examining the mechanism of mammalian initiation. It is important to note that the assay procedures described in this chapter have been developed in the rabbit reticulocyte cell-free system and differences may be found when other mammalian systems are used.

VII. SUMMARY

A soluble fraction containing initiation factors has been obtained from rabbit reticulocytes by washing crude polysomes with 0.5 M KCl. This ribosomal wash fraction has been separated into at least three specific mammalian initiation factors, M_1, M_2, and M_3. M_1 and M_2, partially purified by DEAE-cellulose and Sephadex G-150 chromatography, are reasonably stable, and assays for these factors have been developed which permit studies on the mechanism of mammalian initiation in model systems with both Met-tRNA and Phe-tRNA (aminoacyl-tRNA binding and puromycin peptide release). M_3, required strictly for natural (hemoglobin) messenger translation in the reticulocyte cell-free system, may be the counterpart of F_3 in the bacterial system. This factor has been rather unstable and difficult to study. Indications that factors involved in the initiation process may soon be described from other eukaryotic sources can be found in several recent reports in the literature [40—45].

REFERENCES

[1] R. L. Miller and R. Schweet, Arch. Biochem. Biophys., 125 632 (1968).

[2] J. M. Gilbert and W. F. Anderson, J. Biol. Chem., 245, 2342 (1970).

[3] B, B. Cohen, Biochem. J., 110, 231 (1968).

[4] B. B. Cohen, Biochem. J., 115, 523 (1969).

[5] D. A. Shafritz, P. M. Prichard, J. M. Gilbert, and W. F. Anderson, Biochem. Biophys. Res. Commun., 38, 721 (1970).

[6] P. M. Prichard, J. M. Gilbert, D. A. Shafritz, and W. F. Anderson, Nature, 226, 511 (1970).

[7] D. A. Shafritz and W. F. Anderson, J. Biol. Chem., 245, 5553 (1970).

[8] C. T. Caskey, B. Redfield, and H. Weissbach, Arch. Biochem. Biophys., 120, 119 (1967).

[9] D. A. Shafritz and W. F. Anderson, Nature, 227, 918 (1970).

[10] S. M. Haywood, Proc. Nat. Acad. Sci. USA, 67, 1782 (1970).

[11] A. E. Smith and K. A. Marcker, Nature, 226, 607 (1970).

[12] J. C. Brown and A. E. Smith, Nature, 226, 610 (1970).

[13] K. Takeishi, T. Sekiya, and T. Ukita, Biochim. Biophys. Acta, 199, 559 (1970).

[14] S. Bhaduri, N. K. Chatterjee, K. K. Bose, and N. K. Gupta, Biochem. Biophys. Res. Commun., 40, 402 (1970).

[15] D. Housman, M. Jacobs-Lorena, U. L. Raj Bhandary, and H. F. Lodish, Nature, 227, 913 (1970).

[16] D. B. Wilson and H. Dintzis, Proc. Nat. Acad. Sci. USA, 66, 1282 (1970).

[17] R. Jackson and T. Hunter, Nature, 227, 672 (1970).

[18] A. Yoshida, S. Watanabe, and J. Morris, Proc. Nat. Acad. Sci. USA, 67, 1600 (1970).

[19] W. Culp, J. Morrisey, and B. Hardesty, Biochem. Biophys. Res. Commun., 40, 777 (1970).

[20] D. T. Wigle and G. H. Dixon, Nature, 227, 676 (1970).

[21] S. S. Kerwar, C. Spears, and H. Weissbach, Biochem. Biophys. Res. Commun., 41, 78 (1970).

[22] H. O. Moon, J. F. Collins, and E. S. Maxwell, Biochem. Biophys. Res. Commun., 41, 170 (1970).

[23] D. Richter and F. Lipmann, Nature, 227, 1212 (1970).

[24] J. P. Leis and E. B. Keller, Biochem. Biophys. Res. Commun., 40, 416 (1970).

[25] A. Tarrago, O. Monasterio, and J. E. Allende, Biochem. Biophys. Res. Commun., 41, 765 (1970).

[26] A. Marcus, D. P. Weeks, J. P. Leis, and E. B. Keller, Proc. Nat. Acad. Sci. USA, 67, 1681 (1970).

[27] I. M. London, D. Shemin, and D. Rittenberg, J. Biol. Chem., 183, 749 (1950).

[28] J. M. Gilbert and W. F. Anderson, in Methods in Enzymology, Vol. 19 (K. Moldave, ed.), Academic Press, New York, 1971, in press.

[29] P. C. Yang, K. Hamada, and R. Schweet, Arch. Biochem. Biophys., 125, 506 (1968).

[30] W. L. McKeehan and B. Hardesty, J. Biol. Chem., 244, 4330 (1969).

[31] W. F. Anderson, Biochemistry, 8, 3687 (1969).

[32] A. D. Kelmers, H. O. Weeren, J. F. Weiss, R. L. Pearson, M. P. Stulberg, and G. D. Novelli, in Methods in Enzymology, Vol. 19 (K. Moldave, ed.), Academic Press, New York, 1971, in press.

[33] A. L. Haenni and F. Chapeville, Biochim. Biophys. Acta, 114, 135 (1966).

[34] J. Lucas-Lenard and F. Lipmann, Proc. Nat. Acad. Sci. USA, 57, 1050 (1967).

[35] E. A. Peterson and H. A. Sober, in Methods in Enzymology, Vol. 5 (S. P. Colowick and N. O. Kaplan, eds.), Academic Press, New York, 1962, p. 3.

[36] M. Nirenberg and P. Leder, Science, 145, 1399 (1964).

[37] D. A. Shafritz, D. G. Laycock, and W. F. Anderson, Proc. Nat. Acad. Sci. USA, 68, 496 (1971).

[38] P. Leder and H. Bursztyn, Biochem. Biophys. Res. Commun., 25, 233 (1966).

[39] G. A. Bray, Anal. Biochem., 1, 279 (1960).

[40] A. Marcus, J. Biol. Chem., 245, 955 (1970).

[41] A. Marcus, J. Biol. Chem., 245, 962 (1970).

[42] R. Ascione and G. F. Vande Woude, Am. Chem. Soc., 160th Meeting, Biol. Chem., Abstr. 218, 1970.

[43] F. Grummt, Acta Biol. Med. Germ., 24, K55 (1770).

[44] D. P. Leader, I. G. Wool, and J. J. Castles, Proc. Nat. Acad. Sci. USA, 67, 523 (1970.

[45] K. Moldave and E. Gasior, Fed. Proc., 1971, in press.

Chapter 10

PREPARATION AND MODE OF ACTION OF INTERFERON

Hilton B. Levy and Thomas C. Merigan[1]

Laboratory of Viral Diseases
National Institute of Allergy and Infectious Diseases
Bethesda, Maryland
and
Division of Infectious Diseases
Stanford University School of Medicine
Stanford, California

I. INTRODUCTION 295

II. INTERFERON PREPARATION 297

III. PROPERTIES OF INTERFERON 300

IV. PURIFICATION OF INTERFERONS. 303

V. MODE OF ACTION OF INTERFERON 306

REFERENCES 315

I. INTRODUCTION

In 1957 Isaacs and Lindenmann reported that cells infected with viruses produce a protein, interferon, that when applied to other cells of the same animal species causes these second cells to develop resistance to subsequent infection by the same or different viruses [33]. Interferons have been demonstrated in a wide variety of animal species. This chapter reviews (1) the efforts that have been made in attempting to purify interferon, without presenting detailed procedures; and (2) the studies that have led to some understanding of the molecular basis of the antiviral action of interferon.

Interferon has yet to be obtained in a pure form despite 12 years of effort by biochemists in at least 12 countries. The critical problems are that the molecule has an extremely high biological activity and is produced in minute quantities under conditions in which it is associated with many other protein contaminants (e.g., tissue extracts, serum, chorioallantoic fluid, and so on). All studies made of the mechanism of action have involved interferons of varying degrees of impurity. The contaminating proteins may have had biological or biochemical activity. For example, early reports of the effects of interferon on oxidative metabolism were likely to have been related to contaminating proteins in the preparations [1]. Other sources of antiviral activities, including materials acting by direct viral binding, have been demonstrated in interferon preparations [2]. It has become necessary to develop criteria to define an activity as being attributable to interferons in a preparation. These criteria are: the active material is not destroyed by remaining at pH 2 in the cold for several days; it is not sedimentable at 100,000 X g for 4 hr; it is animal species-specific; it is destroyed by trypsin; it requires preincubation with cells to bring about its effect; its action is blocked by treatment of the cells with actinomycin

D; and, the activity is demonstrable in interferon preparations made in a variety of tissues, both in vivo and in tissue culture, and with different inducers. If possible, parallel dose—response curves should be demonstrated between the effect under study and the antiviral action of the interferon preparation.

All assays for interferon depend on its ability to reduce the capacity of cells to support the growth of viruses. One unit of interferon is generally defined as the amount of interferon that will reduce the yield, or the number of plaques, produced by a given challenge of virus by about 50%. Since test systems vary widely in their sensitivity, a unit in one laboratory may bear no relationship to a unit in another. For a discussion of interferon assays, the reader is referred to the recent symposium on standardization of interferon and interferon inducers [6].

II. INTERFERON PREPARATION

Interferons have been prepared in a wide variety of vertebrate species, using both viral and nonviral inducers in vivo and in vitro. It is quite beyond the scope of this chapter to summarize all these studies, but the reader is referred to several recent reviews on the subject of interferon in which pertinent discussions of interferon preparation are made [3—6]. For the purpose of this chapter, only interferon preparations from tissue culture and eggs are considered, as all in vivo preparations are contained in more crude starting material, hence are subject to more difficulties in purification. Several strains of influenza, acting in 9- to 11-day chick embryos, have provided most of the interferon-rich chick interferon preparations used as starting material for purification by several workers [1,7—9]. The methods of these investigators have formed the basis for most of the schemes evolved for other types of interferons. In each inducer—cell culture system, the kinetics of interferon production are first established, as is the amount of virus or other stimulatory agent required for maximum interferon production. For example, a favorite cell line for the preparation of human interferon in one investigator's laboratory is primary human skin cells derived from foreskin material. Preparation of these monolayers has been described in detail elsewhere [10]. One standard technique is to add between 1 and 5 X 10^7 plaque-forming units of Newcastle disease virus to confluent human fibroblast monolayers growing in 32-oz prescription bottles. The interferon is produced over 48 hr in 20 ml of minimal Eagle's medium (MEM) with antibiotics. The monolayers are incubated in 5% CO_2 at 37°C. Alternatively, the synthetic RNA homopolymer pair, polyinosinic acid (poly I) annealed to polycytidylic acid (poly C), can be used in this same type of monolayer by the addition of 400 μg in 20 ml of MEM to a confluent monolayer growing in a 32-oz bottle and collection of the interferon-containing supernatant at 18—24 hr. However, in certain cell systems the cells must be pretreated with DEAE-dextran for the double-stranded nucleic acid to be active.

For the preparation of interferon to be used in a given animal species, the tissue of choice is generally monolayers of primary cells from that species. Interferon action is rather specific for the animal species in which it is made [11,12], but some exceptions have been noted [13]. In a few animal species, continuous cell lines have been found which are reasonably good sources of virus-stimulated interferon [e.g., L929 cells (mouse), VERO cells (monkey), and RK13 cells (rabbit)].

Explants of cells studied *in vitro* but not grown *in vitro*, such as human white blood cells or mouse or rabbit peritoneal macrophages, are also useful sources of interferon. Unfortunately, such cell preparations often do not respond to inducers in the absence of serum. Hence a more crude interferon preparation results when they are employed, in contrast to tissue culture monolayers which often produce interferon in the absence of serum.

A further problem with a virus-stimulated interferon preparation is inactivation of any residual live virus. This is usually achieved by lowering the pH of the preparation to 2. Interferon is stable under these conditions, and the vast majority of all viruses are unstable. Reovirus withstands this pH, and in this case antibody must be used to inactivate virus infectivity. Ultracentrifugation is a useful technique if only small amounts of virus are present, but it can only be expected to remove 3 logs of virus and this is not sufficient in most situations.

The commonly employed nonviral interferon stimulator is the double-stranded nucleic acid, poly C annealed to poly I, which was discussed above in regard to interferon. This homopolymer pair is active in a wide variety of cell systems in stimulating interferon production. However, in certain systems the cells must be pretreated with DEAE-dextran for the double-stranded nucleic acid to be active. Uninfected or unstimulated cell supernatants have been used as a control for interferon preparations in many studies, although it is clear that these are not sufficiently rigorous controls because virus infection and perhaps even double-stranded nucleic acids release many components from cells not present in uninfected cell supernatants or in cell extracts.

III. PROPERTIES OF INTERFERON

As discussed in Section I, all presently available interferon preparations are impure. Their biological activity has been associated with and is dependent upon a protein. This can be easily shown since incubation of interferon-containing preparations with proteases destroys biological activity, whereas incubation with DNase, RNase, and neuraminidase does not. Variations in the molecular weight, electric charge, and stability have been reported among various interferon preparations (see Ref. 3). The presence of various molecular weight species in a single interferon preparation has been reported on several occasions [14—18]. It has been suggested that the differing molecular sizes in a single preparation of interferon might reflect differences in the cell types producing it [19]. High-molecular-weight species of interferons (greater than 60,000 molecular weight) have only infrequently been reported as being produced in tissue culture, in which most interferons have molecular weights of approximately 40,000—25,000. This is in contrast to activity associated with material found *in vivo*, in which molecular weights range up to 160,000.

In addition to discrete multiple molecular weight species of interferon, a microheterogeneity of interferon has been reported. Studies by Fantes [20] clearly showed that chick interferon peaks, purified by column chromatography, are often wider than expected. Leading edges of such peaks have isoelectric points different from the midpoint or trailing edges. Bodo [21], Stancek et al. [22], and Merigan [23] confirmed this microheterogeneity, working with chick, mouse, and human interferons, respectively. Schonne and associates [24] recently reported that treatment of interferon preparations with neuraminidase from *Clostridium perfringens* decreased this heterogeneity, producing a more homogeneous preparation of rabbit tissue culture interferon. These investigators suggest that the microheterogeneity may be attributable to sialic acid residues and that interferon might be a glycoprotein. Periodate oxidation studies by Gresser and others [25] have also suggested the presence of sugars in interferon. Schonne also noted that pervaporation causes low-molecular-weight rabbit interferon to aggregate, in part,to a high-molecular-weight species which is disaggregated by 4 M urea [26]. Most recently, Carter [27] has reported both mouse and human interferon have a dimeric subunit structure which is converted in low salt to monomers with molecular weights of 19,000 and 12,000, respectively. These latter findings were obtained when partially purified, viral-induced, tissue culture interferon was studied by isoelectric focusing in polyacrylamide—ampholite gels.

In regard to the isoelectric point of interferon preparations, most early studies [1,7,8] indicated chick interferon to have an isoelectric point near neutrality, and mouse interferon [7] was later shown to have a similar isoelectric point. In regard to human interferon, as with chick interferon initially, there were some difficulties involving interaction of the migrating interferon molecules with the supporting media during electrophoresis. Isoelectric points near neutrality were reported [28]. More recent investigations of this problem by Fantes [5] clearly indicated that human interferon has a lower isoelectric point than mouse or chick interferon.

This was shown by both the position of elution during gradient ion-exchange column chromatography and by isoelectric focusing in sucrose—ampholite solutions. Stancek and co-workers have also reached similar conclusions regarding the lower isoelectric point of human interferon as compared to chick or mouse interferons [29]. Later work [23], using isoelectric focusing, also confirmed Fantes' finding of a lower isoelectric point for human interferon. Rabbit interferon was also observed to have an isoelectric point near neutrality by several workers (see review by Fantes [3]).

IV. PURIFICATION OF INTERFERONS

A wide variety of standard chemical techniques used for protein purification has been applied to interferon preparations with success. These methods have included: salting-out procedures; precipitation with trichloroacetic acid, zinc acetate, and ethanol; adsorption to various inert materials such as glass beads or micronized sodium-aluminum silicates; chromatography on ion-exchange resins, including both cellulose and Sephadex types; Sephadex gel filtration; dialysis and zone electrophoresis; and density gradient ultracentrifugation.

Most recently, several workers applied isoelectric focusing in sucrose or acrylamide—ampholite mixtures with relatively good results. This work has been reviewed by Fantes [3]. Despite increases in specific activity up to 40,000-fold as reported by some workers, it is clear that all final products are still contaminated with inert, noninterferon, protein material. The lack of available techniques for complete purification of interferon and the relatively small amounts present that are responsible for the biological activity are well demonstrated by two recent attempts to produce radioactively labeled interferon. Both Paucker and associates [30] and Yamazaki and Wagner [31] failed to demonstrate the incorporation of radioactivity into the interferon molecule using double-labeling experiments with high-specific-activity amino acids applied to normal cells and cells stimulated to produce interferon. The former group, working with L cells, attempted to produce radiolabeled mouse interferon, and the latter group, working with rabbit kidney cells, attempted to produce radio-labeled rabbit interferon.

A relatively simple method of interferon preparation that we have applied to many interferons utilizes molecular sieve chromatography employing dextran gels (in particular G-100 Sephadex chromatography) with 0.1 M sodium phosphate (pH 6.0) buffer. This method has been described in detail [32] and is still recommended as one general approach to interferon purification. It yields good results with a variety of interferon preparations. Low concentrations of protein, for example, 0.05 mg/ml of bovine serum albumin, are often added to column chromatography fractions to stabilize the biological activity of partially purified interferon. Chromatography should be performed under relatively bacteria-free conditions. Samples are kept at 4°C during assay to avoid decreases in biological activity produced by freezing and thawing of purified interferon preparations.

A second method recommended for general use in purification of interferon species is chromatography on carboxymethyl Sephadex. This method has been applied to chick interferon [7], mouse interferon [7], human interferon [5], and rabbit interferon [31]. The general method for purification of each of these interferons begins with adsorption of the interferon-containing preparation, from a prior dextran gel filtration purification, onto carboxymethyl Sephadex (G-25) at pH 4.0; then the interferon activity is eluted by a rising sodium phosphate gradient beginning at 4.0 and reaching pH 8.5 in the final column fractions. Alternatively, the dextran gel step can be used after a carboxymethyl Sephadex step if a concentration step with the pooled peak interferon-containing fractions from the first chromatographic step is employed prior to the second chromatography. Ultrafiltration concentration employing a filtration cell and UM-1 filters (Amicon Corporation, Cambridge, Massachusetts) has been convenient and has given good yields in our laboratory with several different types of interferons.

It should be stressed, in reference to both these chromatographic techniques, that both the molecular heterogeneity and microheterogeneity of interferon species cause a spreading of the interferon activity over a large number of the eluted fractions. However, peak interferon-containing fractions are highly enriched in specific activity and can be quite useful for further studies involving mechanism of action, and so on. It is useful to take the same precautions against bacterial contamination and nonspecific adsorption to glass suggested for the G-100 chromatography effluent when preparing carboxymethyl Sephadex chromatography fractions.

Once partially purified interferon preparations have been obtained, they should be frozen in a solution containing 0.05% serum albumin and kept at -20°C, or lyophilized together with the same inert protein and kept in sealed glass ampuls at -20°C or colder. Unfortunately, even under these conditions, occasionally inexplicable decreases in biological activity have been reported after several months of apparent stability [28].

V. MODE OF ACTION OF INTERFERON

Ever since the original report of Isaacs and Lindenmann [33] that cells infected with viruses respond by synthesizing a protein, interferon, that inhibits the replication of the virus without being grossly toxic to cells, there has been great interest in the mechanism by which this material can exercise this sharp distinction. Some early reports implicated uncoupling of oxidative phosphorylation in this selective viral inhibition, but this was soon ruled out [34]. Inhibition of cellular macromolecule synthesis by interferon preparations was also reported [34—37], but this could be attributed to impurities in the preparation. It was shown that cell growth [38] and gross cell macromolecule synthesis [39] were unaffected by levels of interferon that could induce strong antiviral effects.

Subsequent developments involved two approaches: (1) attempts to understand what changes took place in cells on treatment with interferon that led to a block in virus replication, and (2) attempts to determine which step in virus replication was the primary target of the interferon system. Recently, these two approaches have reunited, and these developments are outlined in the following discussion.

In regard to induction of the antiviral state in cells exposed to interferon, it was early observed that cells treated with interferon required incubation at 37°C for several hours before developing full resistance to virus growth [40]. Incubation at 4°C was only minimally effective [40,41]. It was generally agreed that this meant that metabolic activity was needed for the manifestation of the antiviral action of interferon, and that this action was therefore intracellular.

Taylor [42] noted that actinomycin D blocked the development of antiviral activity in cells treated with interferon and concluded that DNA-directed RNA synthesis was needed for this. She also noted, in an undocumented statement, that fluorophenylalanine (FPA), an inhibitor of protein synthesis, also prevented the antiviral action of interferon. The documentation subsequently presented [43] showed that FPA inhibited the development of interferon action but not when interferon was added 5 hr before FPA. These findings are in accord with those reported independently by Lockart [44] and Levine [45].

These data are consistent with the hypothesis that the direct effect of interferon is derepression of a host genome action, leading to the formation of a new mRNA and the protein for which this mRNA is encoded. On the basis of this concept, the new protein is the effective antiviral substance. It should be noted that the available data also allow the hypothesis that the antiviral substance is an RNA. According to the latter interpretation, treatment of a cell with interferon would first augment the synthesis of a protein that would derepress production of antiviral RNA in the host cell. Neither of these hypothesized substances has been demonstrated to be the actual antiviral substance, however, and it is conceivable that they in turn lead to the formation of further substances. The advantage to an animal in having an antiviral substance, distinct from the interferon molecule, would be that very little interferon from an infected cell could induce a large quantity of an antiviral substance in a second cell. A great amplification of the protective action of interferon could thus be obtained. That there is a need for amplification is indicated by the observations that very little if any interferon is taken up from the medium when it renders tissue culture cells resistant to virus growth [46,47].

Interferon affects an early event in virus replication subsequent to penetration and uncoating of the virus [48,49, 54—57]. The synthesis of new viral nucleic acid is inhibited in interferon-treated cells [50—53]. One very early manifestation of a virus infection is a strong inhibition of cellular RNA and protein synthesis induced in cells by infection with some viruses such as mengovirus or poliovirus. Within 30 sec after infection, such inhibition can be readily detected [58]. In cells pretreated with moderate amounts of interferon, the onset of such inhibition can be delayed [51], and with larger amounts of interferon, even prevented [59]. Under the latter conditions even the cytopathic action of the virus can be minimized. The synthesis of viral proteins in cells infected with Semliki Forest virus under conditions in which no new viral RNA is made is strongly inhibited [60]. Almost every manifestation of virus infection examined has been shown to be inhibited in interferon-treated cells. Some of the events investigated, such as the inhibition of cell macromolecule synthesis by certain RNA-containing viruses, occurred so early in infection as to suggest that the locus of interferon action with these viruses might be an event that involved the input virus particle. This proved to be the case. Through the use of purified mengovirus, labeled with tritiated uridine in its RNA, it was shown that the first association of the viral RNA with a cellular component led to the formation of a particle that sedimented at 50 S in zonal centrifugation in sucrose gradients [61]. It was suggested that this 50 S particle represents an association of the viral RNA with the 40 S ribosomal subunit, but this relationship has never been established. An association of rapidly labeled RNA with subribosomal particles was observed with mRNA in liver cells [64], in HeLa cells [65], with mRNA generated by vaccinia virus in HeLa cells [66], and with RNA of the bacteriophage MS2 [67]. This association is detectable 30 min after infection with mengovirus RNA. Kinetic analysis revealed that from this 50 S particle the viral RNA was complexed with the viral polysome, sedimenting at 250 S. As virus replication proceeds, the RNA of the parental virus is found increasingly on the viral polysome where viral proteins are made. Some viral RNA synthetase is found in association with this polysome. In cells pretreated with interferon, the association of the viral RNA with the 40 S subunit is greatly inhibited. Some virus polysomelike material is found late in infection in these interferon-treated cells, but it is not functional in that it does not synthesize viral protein [61]. As expected, viral RNA synthetase is found in greatly decreased amounts in interferon-treated cells [61—63].

Interferon treatment of vaccinia virus-infected L cells also prevents the incorporation of virus mRNA into a virus polysome [68]. Analysis of this reaction is complicated by the fact that initial association of the mRNA with the 40 S subunit is not always measurable in L cells. To examine the effects of interferon on vaccinia virus mRNA—40 S subunit association it is necessary to utilize HeLa cells in which this association is more readily observable; however, HeLa cells are relatively resistant to interferon.

Further evidence that the interferon-induced virus-resistant state involves the infecting virion comes from studies using Semliki Forest virus with labeled RNA. Early in normal infection it can be shown that this labeled RNA becomes incorporated into double-stranded forms (replicative form and replicative intermediate), these forms being the result of an early copying, by the virus-induced polymerase, of the input genome and immediate association of the newly formed minus strand with the labeled input plus strand. This early event is also blocked in cells rendered resistant to virus by interferon [69].

The failure of input mengovirus genome to form the early association with what is presumed to be the 40 S ribosomal subunit could result either from a modification of the ribosomal subunit or from a modification of the viral RNA occurring upon its entry into the cell that prevents its subsequent attachment. Experiments to distinguish between these two possibilities were made in several laboratories. Interferon-type and control ribosomes from mouse L cells bind cell RNA and bind and translate synthetic polynucleotides equally well, but only the control ribosomes bind and translate mengovirus RNA [70,71]. Also, the interferon-type ribosomes from chick cells do not bind Sindbis virus RNA as well as control ribosomes and translate it into protein even more poorly [72]. A short treatment of the interferon-type chick ribosomes with trypsin augments somewhat their ability to bind and translate the viral RNA, suggesting that the interferon-type ribosomes contain additional proteins that enable them to distinguish between cell and viral RNA [72].

Recently, methods have been developed for synthesizing proteins directed by encephalomycarditis virus RNA with ribosomes and other fractions from mouse cells. The proteins so synthesized have been characterized as virus coat proteins as well as other virus-directed proteins. Ribosome preparations from cells pretreated with interferon synthesize the same proteins as do the control ribosomes but do so less efficiently. Both preparations are stimulated comparably by synthetic mRNAs such as polyuridylic acid [73]. The investigators emphasized that they were dealing with ribosome preparations that contained several known factors in addition to the ribosomes, and possibly one or more unknown factors. Under the circumstances it might be best to ascribe the decreased translation of viral RNA to interferon-induced alterations in ribosomes or factors that associate with ribosomes during translation of mRNAs.

Evidence of a different type to support the suggestion that ribosomes can distinguish between cell and viral mRNA comes from work with the T antigen associated with SV40 virus [74]. Interferon inhibits the synthesis of T antigen in cells productively infected with SV40 virus, in which the virus genome and its mRNA exist as free entities in the cell. In cells transformed by SV40 virus, in which the virus genome may be integrated with the host genome, synthesis of the SV40 T antigen is not blocked by interferon. Thus, in cells in which the SV40 genome is integrated, the mRNA for T antigen is recognized as host mRNA.

It is apparent that the basic effect of interferon is on cells and not on the infecting virus. It is recalled that the establishment of the virus-resistant state in interferon-treated cells requires the synthesis of cellular RNA and protein subsequent to the exposure to interferon and prior to exposure to the challenge virus. Attempts to detect altered proteins in ribosomes from interferon-treated cells that might explain the ability of the ribosomes to distinguish between cellular and viral mRNAs have not been successful, although there were poorly reproducible suggestions that such differences may indeed exist [75,76].

At present, the most likely explanation for the antiviral action of interferon is that interferon treatment of cells leads to the development of a modification in ribosomes or ribosome-associated factors that results in a selective inhibition of translation of viral RNA. What is less clear is the precise point of this inhibition. Since the incorporation of viral mRNA into a ribosomal subunit and a polysome is the point of convergence in the replication of DNA and RNA viruses, it is possible to explain the action of interferon on both types of virus with a single hypothesis.

The development of the virus-resistant state in mouse embryo and chick embryo cells can be demonstrated within 1—4 hr after the initial exposure of the cells to the interferon, and resistance begins to decrease approximately 6—10 hr after removal of the interferon. This is too short a time for any significant replacement of the ribosomes in the cell with new ones. It may be assumed, therefore, that modification of the ribosomes occurs through the addition of an accessory factor or a loosely bound protein rather than a core protein associated with the RNA from its initial time of synthesis.

If it is hypothesized that the interferon-induced antiviral protein is a newly made factor, attached to ribosomes, which is concerned with binding and translation, most of the data obtained to date would be explained. This factor would need to have a specificity that would allow selective inhibition of viral RNA binding and initiation as contrasted with cellular RNA. It would in some ways resemble the sigma factors reported recently, which affect transcription of DNA template by binding to core DNA-dependent RNA polymerase and thus modify the ability of this protein to bind and transcribe. This type of selective association with different types of RNA represents a type of control of protein synthesis at the ribosomal level.

In this regard, the action of interferon resembles that of certain hormones. Isolated ribosomes from hypophysectomized or thyroidectomized rats do not translate added mRNA as well as do ribosomes from normal rats [77]. Ribosomes from diabetic animals also have a low endogenous rate of protein synthesis *in vitro* [78]. Other similarities to hormonal systems have also been pointed out [79].

NOTES

[1] Supported by a grant from the USPHS (Al-05629)

REFERENCES

[1] G. P. Lampson, A. A. Tytell, M. M. Nemes, and M. R. Hilleman, *Proc. Soc. Exptl. Biol. Med.*, *121*, 468 (1963).

[2] C. E. Buckler and S. Baron, *J. Bacteriol.*, *91*, 231 (1966).

[3] K. H. Fantes, in *Interferons, Proceedings of a Symposium Sponsored by the New York Heart Association* (J. Vilcek, ed.), Little, Brown, Boston, 1970, p. 113.

[4] K. H. Fantes, in _Interferon and Interferon Inducers_ (N. Finter, ed.), Saunders, Philadelphia, 1971, in press.

[5] K. H. Fantes, _Ann. N. Y. Acad. Sci._, _173_, 118 (1970)

[6] _Symposium Series on Immunobiological Standardization. International Symposium on Interferon and Interferon Inducers_ (F. Perkins, ed.), Karger, Basel, 1970.

[7] T. C. Merigan, C. A. Winget, and C. B. Dixon, _J. Mol. Biol._, _13_, 679 (1965).

[8] K. H. Fantes, _Nature_, _207_, 1298 (1965).

[9] G. Bodo, _Monatsh. Chem._, _99_, 1 (1968).

[10] T. C. Merigan, in _Methods for Study of Cellular Immunity_ (B. Bloom, K. Hirschorn, and H. S. Lawrence, eds.), Academic Press, New York, 1971.

[11] T. C. Merigan, _Science_, _145_, 811 (1964).

[12] S. Baron and C. Buckler, _Science_, _145_, 813 (1964).

[13] J. Desmyter, W. E. Rawls, and J. H. Melnick, _Proc. Natl. Acad. Sci._, _59_, 69 (1968).

[14] G. P. Lampson, A. A. Tytell, M. M. Nemes, and M. R. Hilleman, _Proc. Soc. Exptl. Biol. Med._, _121_, 377 (1966).

[15] Y. Ke, T. C. Merigan, and M. Ho, _Nature_, _211_, 541 (1966).

[16] T. C. Merigan and W. J. Kleinschmidt, _Nature_, _212_, 1383 (1966).

[17] J. V. Hallum, J. S. Youngner, and W. R. Stinebring, _Bacteriol. Proc._, p. 118 (1966).

[18] E. Falcoff, F. Fournier, and C. Chany, Ann. Inst. Pasteur, 111, 241 (1966).

[19] T. C. Merigan, Bacteriol. Rev., 31, 138 (1967).

[20] K. H. Fantes, Science, 163, 1198 (1969).

[21] G. Bodo, personal communication, 1969.

[22] D. Stancek, R. R. Golgher, and K. Paucker, in Interferons, Proceedings of a Symposium Sponsored by the New York Heart Association (J. Vilcek, ed.), Little, Brown, Boston, 1970, p. 134.

[23] T. C. Merigan, personal communication, 1969.

[24] E. Schonne, A. Billiau, and P. DeSomer, In Symposium Series on Immunobiological Standardization. International Symposium on Interferon and Interferon Inducers (F. Perkins, ed.), Karger, Basel, 1970.

[25] I. Gresser, in Symposium Series on Immunobiological Standardization. International Symposium on Interferon and Interferon Inducers (F. Perkins, ed.), Karger, Basel, 1970, p. 70.

[26] E. Schonne, A. Billiau, and P. DeSomer, in Proceedings of an International Symposium on Interferon (C. Chany, ed.), L'Institut National de la Sante et de la Recherche Medicale, Paris, 1970, in press.

[27] W. Carter, Proc. Natl. Acad. Sci., 67, 620 (1970).

[28] T. C. Merigan, D. Gregory, and J. Petralli, Virology, 20, 515 (1966).

[29] D. Stancek, M. Gressnerova, and K. Paucker, Virology, 41, 740 (1970).

[30] K. Paucker, B. J. Berman, R. R. Golger, and D. Stancek, J. Virol., 5, 145 (1970).

[31] S. Yamazaki and R. Wagner, J. Virol., 5, 270 (1970).

[32] T. C. Merigan, in Fundamental Techniques in Virology (K. Habel and N. Salzman, eds.), Academic Press, New York, 1969, Chapter 33.

[33] A. Isaacs and J. Lindenmann, Proc. Roy. Soc. Ser. B., 147, 248 (1957).

[34] H. B. Levy, L. F. Snellbaker, and S. Baron, Virology, 21, 48 (1963).

[35] C. Cocito, E. DeMaeyer, and P. DeSomer, Life Sci., 1, 753 (1962).

[36] C. Cocito, E. Schonne, and P. DeSomer, personal communication, 1970.

[37] J. A. Sonnabend, Nature, 203, 496 (1964).

[38] S. Baron, T. C. Merigan, and M. L. McKerlie, Proc. Soc. Exptl. Biol. Med., 121, 53 (1966).

[39] H. B. Levy and T. C. Merigan, Proc. Soc. Exptl. Biol. Med., 121, 53 (1966).

[40] J. Lindenmann, D. C. Burke, and A. Isaacs, Brit. J. Exptl. Pathol., 38, 551 (1957).

[41] J. Vilcek and B. Rada, Acta Virol., 6, 9 (1962).

[42] J. Taylor, Biochem. Biophys. Res. Commun., 14, 447 (1964).

[43] R. M. Friedman and J. A. Sonnabend, Nature, 203, 366 (1964).

[44] R. Z. Lockart, Jr., Biochem. Biophys. Res. Commun., 15, 513 (1964).

[45] S. Levine, Virology, 24, 229 (1964).

[46] C. E. Buckler, S. Baron, and H. B. Levy, Science, 152, 80 (1966).

[47] J. S. Youngner and W. R. Stinebring, personal communication, 1970.

[48] S. E. Grossberg and J. J. Holland, J. Immunol., 88, 708 (1962).

[49] P. DeSomer, A. Prinzie, P. Denys, Jr., and E. Schonne, Virology, 16, 62 (1962).

[50] M. Ho, Proc. Soc. Exptl. Biol. Med., 112, 511 (1963).

[51] H. B. Levy, Virology, 22, 575 (1964).

[52] R. M. Friedman and J. Sonnabend, Nature, 206, 532 (1965).

[53] I. Gordon, S. Chenault, D. Stevenson, and J. Acton, J. Bacteriol., 91, 1230 (1966).

[54] S. N. Gosh and G. E. Gifford, Virology, 27, 186 (1965).

[55] S. Ohno and T. Nozima, personal communication, 1964.

[56] S. Baron and H. B. Levy, Ann. Rev. Microbiol., 20, 291 (1966).

[57] H. B. Levy, S. Baron, and C. E. Buckler, in Biochemistry of Interferon (H. B. Levy, ed.), Dekker, New York, 1969, p. 579.

[58] R. Franklin and D. Baltimore, in Viruses, Nucleic Acids and Cancer, Univ. of Texas M. D. Anderson Hospital and Tumor Institute, Williams & Wilkins, Baltimore, Maryland, 1964, p. 301.

[59] A. T. Haase, S. Baron, H. Levy, and J. A. Kasel, J. Virol., 4, 490 (1969).

[60] R. M. Friedman, J. Virol., 2, 1081 (1968).

[61] H. B. Levy and W. A. Carter, J. Mol. Biol., 31, 561 (1968).

[62] N. Miner, W. J. Ray, Jr., and E. H. Simon, Biochem. Biophys. Res. Commun., 24, 264 (1966).

[63] J. A. Sonnabend, E. M. Martin, E. Mecs, and K. H. Fantes, J. Gen. Virol., 1, 31 (1967).

[64] E. Henshaw, M. Revel, and H. Hyatt, J. Mol. Biol., 14, 241 (1965).

[65] M. Girard, H. Latham, S. Penman, and J. E. Darnell, J. Mol. Biol., 11, 187 (1965).

[66] W. K. Joklik and Y. Becker, J. Mol. Biol., 13, 511 (1965).

[67] G. N. Godson and R. J. Sinsheimer, J. Mol. Biol., 23, 495 (1967).

[68] W. K. Joklik and T. C. Merigan, Proc. Natl. Acad. Sci. U.S., 56, 558 (1966).

[69] R. M. Friedman, K. H. Fantes, H. B. Levy, and W. A. Carter, J. Virol., 1, 1168 (1967).

[70] W. A. Carter and H. B. Levy, Science, 155, 1254 (1967).

[71] W. A. Carter and H. B. Levy, Biochem. Biophys. Acta, 155, 437 (1968).

[72] P. I. Marcus and J. M. Salb, in The Interferons (G. Rita, ed.), Academic Press, New York, 1968, p. 111.

[73] I. Kerr, E. M. Martin, and P. Dobos, Proc. Tenth Intern. Congr. Microbiology, 1970, p. 168, and personal communication, 1970.

[74] M. N. Oxman, S. Baron, P. H. Black, K. K. Takemoto, K. Habel, and W. P. Rowe, Virology, 32, 122 (1967).

[75] H. B. Levy, unpublished observations, 1970.

[76] I. M. Kerr, E. M. Martin, J. A. Sonnabend, and D. M. Metz, Biochem. J., 43, 44 (1969).

[77] L. D. Garren, R. L. Ney, and W. W. Davis, Proc. Natl. Acad. Sci. U.S., 53, 1443 (1965).

[78] I. G. Wool and P. Cavicchi, Biochemistry, 6, 1230 (1967).

[79] S. Baron, in Interferons (N. B. Finter, ed.), North Holland, Amsterdam, 1966, p. 268.

AUTHOR INDEX

Numbers in brackets are reference numbers and indicate that an author's work is referred to although his name is not cited in the text. Underlined numbers give the page on which the complete reference is listed.

A

Abi, S., 173[34], 186
Acs, G., 153[15], 173[15], 185
Acton, J., 309[53], 319
Adamson, S.D., 18, 28, 248, 249[23], 251[23], 263
Ahmad-Zadeh, C., 31[4], 55
Allen, E., 35[12], 36[12], 51, 52[12], 53[12,22], 56, 57
Allende, J.E., 143[21], 145, 267[25], 292
Alexandre, Y., 235[16], 262
Anderson, W.F., 62[3], 107[50], 108[3], 112, 115, 266[2,5, 6,7], 267[6,9], 268[28], 272[2,5,7], 274[2,5,7,9,28, 31], 276[5,7], 279[7], 280 [5,6,7], 281[9], 282[37], 284[31], 285[37], 291, 293, 294
App, A., 130[6], 144
Arlinghaus, R., 32[7], 43[7, 14], 55, 56, 62[4], 89[35], 96[39], 105[45], 109[53, 55], 112, 114, 115, 116
Aronson, R.F., 148[3], 184
Ascione, R., 32[7], 43[7], 55, 62[4], 83[21], 84[27], 105 [45], 111[57], 112-116, 291[42], 294
Attardi, B., 31[5], 55
Attardi, G., 31[5], 55

B

Bachrach, H.L., 84[29,30], 114
Bailey, P., 2[2], 3[2], 26
Balbinder, E., 215[19], 228
Baliga, B.S., 44[16], 56
Baltimore, D., 83[24], 113, 309[58], 320
Baron, S., 296[2], 299[12], 306[34], 307[34,38], 309[46, 56,57,59], 313[74], 315[79], 315, 316, 318-321
Battaner, E., 21[27], 28
Bauer, G.E., 62[5], 112
Becker, Y., 63[12], 83[22], 113, 159[26], 186, 310[66], 320
Beecher, G.R., 47[19], 56
Beisson, J., 226[27], 229
Benrubi-Dumont, M., 235[16], 262
Berman, B.J., 304[30], 318
Bewley, J.P., 136[15], 142[18], 145
Bhaduri, S., 267[14], 292
Bickle, A., 31[4], 55
Billiau, A., 301[24], 302[26], 317
Bishop, J.O., 83[25], 105[46], 109[53], 114, 115, 116, 214[15], 215[16], 216[15], 228, 236[17], 262
Black, P.H., 313[74], 321
Blobel, G., 35[11], 48[11], 56, 82[18], 92[38], 93[38], 113, 115, 122[5], 126

Bodo, G., 298[9], 301, 316, 317
Bonanou, S., 182[41], 187
Bondy, S.C., 153[16], 173[16], 185
Bongiorno, M., 44[17], 56, 75[16], 113
Borsook, H., 232[15], 236[18], 251, 262, 263
Borun, T.W., 62[8], 112, 233 [13], 262
Bose, K.K., 267[14], 292
Bray, G.A., 283[39], 294
Britten, R.J., 53, 57
Breillatt, J., 33[10], 56, 73[15], 113
Brot, N., 134[13], 145
Brown, B.J., 148[5,7,8], 155 [7], 162[7], 163[7], 166 [7], 169[7,8], 172[7], 175[7], 182[8], 184
Brown, D.M., 150[13], 153 [18], 155[18], 16][18], 166[13], 185
Brown, J.C., 267[12], 292
Bryer, C.B., 2[4], 26
Buckler, C.E., 296[2], 299[12], 309[46,57], 315, 316, 319
Bullock, G., 4[9], 27
Burke, D.C., 307[40], 318
Burny, A. 232[4,5], 238[4], 239[19], 241[3], 246 [19], 261, 262
Bursztyn, H., 134[12], 144, 282[38], 294

C

Cain, J.C., 150[12], 185
Campagnoni, A.T., 148[6], 167[6], 169[6], 172[6], 184
Campbell, M.K., 148[4], 184
Carter, W.A., 302, 310[61], 311[69], 312[70,71], 317, 320
Caskey, C.T., 267[8], 274[8], 291
Castles, J.J., 9[16], 11[17], 27, 291[44], 294
Cavicchi, P., 3[5], 6[10], 16 [10], 26, 27, 315[78], 321
Chappeville, F., 110, 116, 274 [33], 293
Chantrenne, H., 232, 261
Chany, C., 301[18], 317
Chatterjee, N.K., 267[14], 292
Chenault, S., 309[53], 319
Chesters, J.K., 207[11], 228
Ciferri, O., 101[43], 115
Clouet, D.H., 148[2], 184
Cocito, C., 307[35,36], 318
Cohen, B.B., 266[3,4], 291
Collins, J.F., 267[22], 292
Cooper, W.K., 47[19], 56
Cordes, E., 53[22], 57
Cottone, M.A., 65[13], 113
Cox, R.A., 182[41], 187
Cravioto, B., 31[5], 55
Culp, W.J., 130[8], 144, 267 [19], 292

D

Darnell, J.E., 30[3], 55, 63 [12], 83[20,22,23,24], 113, 159[26], 177[38], 186, 310 [65], 320
Datta, R.K., 153[17], 185
Davis, W.W., 315[77], 321
Day, L., 223[23], 228
De Carli, L., 101[43], 115
Delius, H., 31[4], 55
De Maeyer, E., 307[35], 318
Denys, P., Jr., 309[49], 319
Desmyter, J., 299[13], 316
De Somer, P., 301[24], 302 [26], 307[35,36], 309[49], 317, 318, 319
Dickman, S.R., 33[10], 56, 67[14], 73[15], 113
Dingman, C.W., 177[35], 186
Dintzis, H., 267[16], 292
Dixon, C.B., 298[7], 302[7], 305[7], 316
Dixon, G.H., 267[20], 292

Dobos, P., 312[73], 321
Dreyfus, J.C., 232[10], 235 [16], 239[10], 262
Dounce, A.L., 65, 113
Dowben, R.M., 3[7], 4[8], 27
Dunn, A.J., 169[30], 172[30], 173[30], 186

E

Egyhazi, E., 177[36], 186
Ekholm, R., 150[10], 184
Eliasson, E., 62[5], 112
Ertel, R., 134[13], 145
Evans, M.J., 232[7], 261, 262

F

Falcoff, E., 301[18], 317
Falvey, A.K., 8[15], 11[15], 27, 105[49], 115, 171 [31], 182[31], 186, 183 [43], 187
Fantes, K.H., 297[4,5], 298 [8], 301, 302[8], 303, 305[5], 310[63], 311[69], 315, 316, 317, 320
Farr, A.L., 171[32], 186
Favelukes, G., 236[17], 262
Finger, I., 215[20], 228
Fisher, E.H., 251[26], 263
Fisher, J.M., 249[25], 252, 263
Fleck, A., 6[11], 27
Florini, J.R., 2[4], 26
Fournier, F., 301[18], 317
Fraenkel-Conrat, H., 136[16], 145
Franklin, R., 309[58], 320
Freedman, M.L., 249[25], 263
Friedman, R.M., 308[43], 309 [52,60], 311[69], 319, 320

G

Gaffey, T.A., 4[8], 27
Gallwitz, D., 232, 262
Garren, L.D., 315[77], 321
Gasior, E., 291[45], 294
Ghosh, J.J., 153[17], 185
Gibbons, I.R., 226[28], 229
Gierer, A., 159[25], 186
Gifford, G.E., 309[54], 319
Gilbert, J.M., 62[3], 107[50], 108[3], 112, 115, 266[2,5, 6], 267[6], 268[28], 272 [2,5], 274[25,28], 276[5], 280[5,6], 285[6], 291, 293
Gilbert, W., 11[18], 27
Girard, M., 83[24], 113, 310 [65], 320
Glassman, E., 53[22], 57
Glowacki, E., 7[13], 27
Godchaux, W. III., 18[24], 28, 264[23], 249[23], 251[23], 263
Godson, G.N., 310[67], 320
Goldberg, B., 62[10], 112
Golgher, R.R., 301[22], 304 [30], 317, 318
Goodwin, F., 183[45], 187
Gordon, I., 309[53], 319
Gordon, J., 18[21], 28
Gosh, S.N., 309[54], 319
Graham, A.F., 62[9], 112
Gray, E.G., 150[4], 185
Green, H., 62[10], 112
Gregory, D., 302[28], 306[28], 317
Gresser, I., 302, 317
Gressnerova, M., 303[29], 318
Grollman, A.P., 136[14], 145
Gross, F., 36[13], 56
Gross, J., 48, 56, 82[19], 113
Gross, P.R., 232, 262
Grossberg, S.E., 309[48], 319
Gruber, C.P., 148[7], 155[7], 162[7], 163[7], 166[7], 169[7], 177[7], 175[7], 184
Grummt, F., 291[43], 294
Gupta, N.K., 267[14], 292

H

Haase, A.T., 309[59], 320
Habel, K., 313[74], 321

Hadjiolov, A.A., 177[39], 186
Haenni, A.L., 110, 116, 274[33], 293
Hall, C.E., 159[24], 185
Hallum, J.V., 301[17], 316
Hamada, K., 270[29], 293
Hardesty, B., 18[22], 28, 130[8,9], 144, 267[19], 292
Haschemeyer, A.E.V., 48, 56, 82[19], 113
Heeter, M., 83[20], 113
Heintz, R., 43[14], 56, 96[39], 115
Heller, C., 21[27], 28, 215[20], 228
Henshae, E.C., 62[7], 75[7], 75[7], 112
Henshaw, E., 310[64], 320
Herbert, E., 18[24], 28, 248[23], 249[23], 251[23], 263
Herbert, W.J., 215, 228
Heywood, S.M., 3[7], 27, 32[8], 55, 232, 262, 267[10], 291
Higginson, B., 182[41], 187
Hill, A.Z., 46[18], 56
Hille, M., 108[51], 115
Hilleman, M.R., 296[1], 298[1], 301[14], 302[1], 315, 316
Hirsch, C.A., 62[7], 75[7], 112
Ho, M., 301[15], 309[50], 316, 319
Hoagland, M.B., 46[18], 56
Holland, J.J., 309[48], 319
Honold, G.R., 47[19], 56
Housman, D., 267[15], 292
Hsu, K.C., 225[262, 229
Huez, G., 232, 239[19], 246[19], 261, 262
Hultin, T., 62[5], 112
Humphrey, R.M., 62[6], 112
Hunt, J.A., 232, 239, 262
Hunter, T., 267[17], 292
Huston, R.L., 47[19], 56
Hyatt, H., 310[64], 320
Hyden, H., 150[10], 177[36, 37], 184, 186

I

Imahori, K., 100[41], 115
Isaacs, A., 295[33], 306[33], 307[40], 318

J

Jackson, R., 159[22], 185, 267[17], 292
Jacob, F., 231, 261
Jacobs-Lorena, M., 267[15], 292
Johnson, H.M., 223, 228
Johnson, T.C., 148[9], 184
Johnston, F.B., 128[2], 144
Joklik, W.K., 310[66], 311[68], 320
Jones, I.G., 214[9], 215[9], 228

K

Kaltreider, H.B., 121[4], 126
Kanagalingam, K., 182[41], 187
Kasel, J.A., 309[59], 320
Ke, Y., 301[15], 316
Kedes, L.H., 232, 262
Keighley, G., 251[16], 263
Keiwar, S.S., 182[44], 187
Keller, E.B., 143[19,20], 145 153[19], 157[20], 185, 267[24,26], 292, 293
Kelmers, A.D., 274[32], 293
Kerr, I., 312[73], 313[76], 321
Kerwar, S.S., 267[21], 292
Kleinschmidt, W.J., 301[16], 316
Knopf, P., 63[11], 112
Korner, A., 2, 26, 100, 115, 158[21], 159[22], 173[33], 185, 186
Kruh, J., 36[13], 56, 232[15, 16], 262
Kuff, E.L., 163[29], 186
Kurihara, K., 2[2], 3[2], 26
Kurland, C. 30[2], 55

L

Labrie, F., 239, 262
Lamfrom, H., 7[13], 27
Lampson, G.P., 296[1], 298 [1], 301[14], 302[1], 315, 316
Lanyon, W.G., 239[22], 263
Last, J., 108[51], 115
Latham, H., 310[65], 320
Laycock, D.G., 282[37], 285 [37], 294
Laycock, R.E., 232, 241[9], 255[9], 262
Lebleu, B., 232[4], 238[4], 239, 261, 262
Leader, D.P., 291[44], 294
Leder, P., 132[11], 134 [12], 144, 280, 282[38], 293, 294
Legocki, A.B., 128[1], 132 [1], 144
Leis, J.P., 143[19,20], 145, 267[24,26], 292, 293
Lerner, M.P., 148[9], 184
Levine, S., 308, 319
Levintow, L., 83[25], 114
Levy, H.B., 306[34], 307[34, 39], 309[46,51,56,57,59], 310[61], 311[69], 312[70, 71], 313[75], 318, 319, 320, 321
Lindenmann, J., 295[33], 306 [33], 307[40], 318
Lingrel, J., 32[6], 37[6], 55, 232[7], 236[18], 241[9], 255[9], 261, 262
Lipmann, F., 18[21], 19[25], 28, 109[52], 116, 190, 227, 267[23], 274[34], 292, 293
Littlefield, J.W., 153[19], 185
Lockard, R.E., 32[6], 37[6], 55, 232, 241[9], 255[9], 262
Lockart, R.Z., Jr., 308, 319
Lodish, H.F., 267[15], 292
Loening, U., 209[13], 228
London, I.M., 268[27], 293
Low, R.B., 2[2,3], 3[2], 7[3], 26, 105[48], 115
Lowry, O.H., 171, 186
Lucas-Lenard, J., 109[52], 116, 274[34], 293
Luckins, A.G., 227[30], 229
Lynch, P.M., 4[8], 27

M

McAllister, H.C., 18[23], 28
Macindoe, H., 190[2], 199[2], 200[2], 207[12], 209[12], 215[18], 227, 228
McKeehan, W.L., 130[9], 144, 293
McKerlie, M.L., 307[38], 318
Macpherson, I., 43[15], 56
Maden, B.E.H., 17[20], 27
Magar, J., 190, 227
Mahler, H.R., 148[4,6,8], 167 [6], 168[8], 169[6,8], 172[6], 182[8], 184
Maizel, J.V., Jr., 83[23,24], 86[31], 87[34], 89[34], 92[37], 113, 114
Maleknia, N., 232[10], 235[16], 239[10], 262
Mans, R.J., 51, 56
Mansbridge, J., 158[21], 185
Marbaix, G., 232[4,5], 238[4], 239[19], 241[3], 246[19], 261, 262
Marchesi, V.T., 226[27], 229
Marcker, K.A., 62[2], 112, 267[11], 292
Marcus, A., 128[1,3,4,5], 132 [1], 136[15], 138[4], 139 [17], 141[18], 143[19], 144, 145, 267[26], 291[40, 41], 293, 294
Marcus, P.I., 312[72], 321
Marks, P., 36[13], 56
Martin, E.M., 310[63], 312[73], 313[76], 320, 321
Martin, T.E., 2[3], 7[3,12], 11 [17], 16[19], 26, 27, 105 [48], 115, 182[42], 187

Mase, K., 163[28], 173[34], 186
Matthaei, J.H., 20[26], 21 [26], 28
Maxwell, E.S., 267[22], 292
Mecs, E., 310[63], 320
Melnick, J.H., 299[13], 316
Merigan, T.C., 298[7,10], 299 [11], 301[15,16,19], 302 [7,]8], 303[23], 304[32], 305[7], 306[28], 307[38, 39], 311[68], 316, 317, 318, 320
Merits, I., 150[12], 185
Metz, D.M., 313[76], 321
Miller, R., 32[9], 37[9], 55, 62[1], 103, 106[44], 108 [1], 103, 106[44], 108 [1], 115, 266[1], 272[1], 274[1], 291
Minard, F.N., 150[12], 185
Miner, N., 310[62], 320
Miyazawa, F., 100[400], 115
Moldave, K., 110[54], 116, 130[7], 144, 291[45], 294
Monasterio, O., 143[21], 145, 267[25], 292
Monod, J., 231, 261
Moon, H.O., 267[22], 292
Moore, W.J., 148[4], 184
Monro, R.E., 17[20], 21[27], 27, 28
Monty, K.J., 65[13], 113
Morimoto, T., 7[14], 27
Morris, J., 267[18], 292
Morrisey, J., 267[12], 292
Mosteller, R.D., 18[22], 28, 130[8], 144
Mudd, J.A., 86[33], 114
Mueller, G.C., 232, 262
Munro, A.J., 159[22], 185
Munro, H.N., 6[11], 27, 44 [16], 56
Murthy, M.R.V., 148[1], 173 [1], 184

N

Neidle, A., 153[15], 173[15], 185
Nemes, M.R., 296[1], 298[1], 301[14], 302[1], 315, 316
Neth, R., 21[27], 28
Ney, R.L., 315[77], 321
Nirenberg, M., 20[26], 21[26], 28, 132[112, 144, 280, 293
Noll, H., 53, 57, 159[27], 160 [27], 161[27], 186
Nomura, M., 30[1], 55
Novelli, G.D., 51, 56, 274[32], 293
Nozima, T., 309[55], 319
Nwagwu, M., 232, 262

O

Ochoa, S., 108[51], 115
Ohno, S., 309[55], 319
Olijnyk, O.R., 100[40], 115
Onorato, F., 215[20], 228
Oxman, M.N., 313[74], 321
Oyer, D., 2[2], 3[2], 26

P

Padieu, P., 235[16], 262
Palade, G.E., 121[3], 126
Parisi, B., 101[43], 115
Pate, S., 65[13], 113
Paucker, K., 301[22], 303[29], 304, 317, 318
Paul, J., 239[22], 263
Pavlovec, A., 105[47], 115
Peache, S., 150[13], 177[40], 179[40], 183[40], 185, 187
Pearson, P., 31[4], 55
Pearson, R.L., 274[32], 293
Penman, S., 63[12], 83[22], 84[28], 113, 114, 159[26], 186, 310[65], 320
Perani, A., 101[43], 115
Perry, S.V., 153[16], 173[16], 185
Peterman, M.L., 105[47], 115
Peterson, E.A., 277[35], 293
Petralli, J., 302[28], 306 [28], 317
Phillips, W.D., 119[1], 126

Polatnick, J., 83[21], 84[30], 89[35], 113, 114
Potter, V.R., 35[11], 48[11], 56, 82[18], 92[38], 93 [38], 113, 115, 122[5], 126
Preer, J.R., 213[14], 215[19], 228
Pressman, D., 223[23], 228
Prevec, L., 62[9], 112
Prichard, P.M., 62[3], 107 [50], 108[3], 112, 115, 266[5,6], 267[6], 272[5], 274[5], 276[5], 280[5,6], 285[6], 291
Prinzie, A., 309[49], 319
Pronczuk, A.W., 44[16], 56
Provost, C., 148[5], 153[18], 155[18], 162[18], 184, 185

R

Rabinowitz, M., 249[25], 252, 263
Rada, B., 307[41], 318
Raj Bhandary, U.L., 267[15], 292
Randall, R.J., 171[32], 186
Rappoport, D.A., 148[1], 153[14], 169[14], 170[14], 173[1,14], 184, 185
Ratner, M., 148[2], 184
Ravel, J.M., 18[22], 28
Rawls, W.E., 299[13], 316
Ray, W.J., Jr., 310[62], 320
Rdzok, E.J., 150[12], 185
Redfield, B., 267[8], 274[8], 291
Redman, C.M., 92[36], 114
Reibel, L., 235[16], 262
Reisner, A.H., 190[2], 199 [2], 200[2], 207, 215[18], 227, 228
Renaud, F.L., 226[28], 229
Revel, M., 310[64], 320
Rich, A., 3[7], 27, 63[11], 112, 159[24], 184
Richter, D., 267[23], 292
Risebrough, R.W., 159[23], 185
Rittenberg, D., 268[27], 293
Robbins, E., 62[8], 112, 232 [13], 262
Roberts, N.E., 163[29], 186
Roberts, R.B., 53, 57
Roberts, S., 148[5,7], 150 [13], 153[14,18], 155[7,18], 162[7,18], 163[7], 166[7, 13], 169[7,14], 170[14], 172[7], 173[14], 175[7], 177[40], 179[40], 183[40], 184-187
Rolleston, F.S., 2[3], 7[3], 26, 105[48], 115
Rosebrough, N.J., 171[32], 186
Roumiantzeff, M., 92[37], 114
Rowe, J., 207[12], 209[12], 215[18], 226[28], 228, 229
Rowe, W.P., 313[74], 321
Rubin, A.L., 119[1], 126, 148 [3], 184
Russell, E., 236[17], 262

S

Salas, M., 108[51], 115
Salb, J.M., 312[72], 321
Sarma, D.S.R., 44[17], 56, 75[16], 113
Sauberlich, H.E., 47[19], 56
Scharff, M., 62[8], 87[34], 89[34], 112, 114, 232[13], 262
Schenkein, I., 225[25], 229
Scherrer, K., 83[22], 113, 159[26], 186
Schonne, E., 301, 302[26], 307[36], 309[49], 317, 318, 319
Schrader, L.E., 47[19], 56
Schram, E., 239[19], 246[19], 262
Schulman, H.M., 249[24], 263
Schweet, R.S., 18[23], 28, 32[9], 35[12], 36[12], 37 [9], 43[14], 51, 52[12], 53[12,22], 55, 56, 57, 77 [17], 96[39], 103, 106[44],

109[53,55], 113, 115, 116, 236[17], 262, 266[1], 270 [29], 272[1], 274[1], 291, 293
Sekiya, T., 267[13], 292
Shaeffer, J.R., 62[6], 109[53, 55], 112, 116
Shafritz, D.A., 62[3], 107[50], 108[3], 112, 115, 187, 266 [5,6,7], 267[6], 272[57], 274[5,7,9], 276[5,7], 279 [7], 280[5,6,7], 281[9], 282[37], 285[6,37], 291, 294
Shapira, G., 232, 235[16], 262
Shemin, D., 268[27], 293
Sidransky, H., 44[17], 56, 75[16], 113
Siekevitz, P., 121[3], 126
Siler, J., 110[54], 116, 130 [7], 144
Simon, E.H., 310[62], 320
Sinden, R.E., 190[5], 201[10], 206[5], 217[22], 224[10, 24], 226[24], 227, 228, 229
Sinozawa, T., 100[41], 115
Sinsheimer, R.J., 310[67], 320
Smith, A.E., 62[2], 112, 267 [11,12], 292
Snellbaker, L.F., 306[34], 307[34], 318
Sober, H.A., 277[35], 293
Soeiro, R., 177[38], 186
Soldo, A.T., 193, 228
Sommerville, J., 190[3-6], 199[3], 206[5], 215[17], 218[6], 221[6], 222[4,6], 226[17], 227, 228
Sonnabend, J.A., 307[37], 308 [43], 309[52], 310[63], 313[76], 318, 319, 320, 321
Sonneborn, T.M., 193, 228
Spears, C., 182[44], 187, 267[21], 292
Sporn, M.B., 177[35], 186
Staehlin, T., 8[15], 11[15], 27, 105[49], 115, 159[27], 160[27], 161[27], 171[31], 18][31], 183[43], 186, 187
Stancek, D., 301, 303[29], 304 [30], 317, 318
Steers, E., 226[27], 229
Stenzel, K.H., 119[1], 126, 148[3], 184
Stern, H., 128[2], 144
Stevenson, D., 309[53], 319
Steward, D.L., 62[6], 112
Stewart, M.L., 136[14], 145
Stinebring, W.R., 301[17], 309[47], 316, 319
Stirewalt, W.S., 2[2], 3[2,5], 9[16], 26, 27
Stoker, M., 43[15], 56
Stulberg, M.P., 274[32], 293
Sugano, H., 163[28], 186
Summers, D.F., 83[23,24,25], 86[33], 87[34], 89[34], 92[37], 113, 114
Szent-Györgyi, A.G., 3[6], 27

T

Takahashi, Y., 163[28], 173 [34], 186
Takeishi, K., 267[13], 292
Takemoto, K.K., 313[74], 321
Talal, N., 120[2], 121[4], 126
Tamaoki, T., 100[40], 115
Tarrago, A., 267[25], 292
Tarrago, H., 143[21], 145
Tashiro, Y., 7[14], 27
Taylor, J., 307, 318
Temmerman, J., 239, 262
Tencheva, Z.S., 177[39], 186
Tewari, S., 148[4], 184
Thireau, A.M., 235[16], 262
Thompson, D.D., 119[1], 126
Tichonicky, L., 235[16], 262
Tilley, C.J., 100[40], 115
Tissieres, A., 31[4], 55, 159 [23], 185
Traut, R., 31[4], 55
Tucker, J.B., 226[29], 229

Tytell, A.A., 296[1], 298[1], 301[14], 302[1], 315, 316

U

Uhr, J.W., 225[25], 229
Ukita, T., 267[13], 292

V

Vande Woude, G.F., 83[21], 84[27], 89[35], 111[57], 113, 114, 116, 291[42], 294
van Venrooij, W.J.M., 62[7], 75[7], 112
Van Wagtendonk, W.J., 193, 228
Vaughan, M.H., 177[38], 186
Vazquez, D., 21[27], 28
Verney, E., 44[17], 56, 75[16], 113
Vickerman, K., 227[30], 229
Vilcek, J., 307[41], 318
von Ehrenstein, G., 19[25], 28

W

Waelsch, H., 153[15], 173[15], 185
Wagner, R., 304, 305[31], 318
Wahba, A., 108[51], 115
Warner, J.R., 63[11], 112, 159[24], 177[38], 185, 186
Watanabe, S., 267[18], 292
Watson, J.D., 159[23], 185
Weeks, D.P., 128[5], 136[15], 139[17], 143[19], 144, 145, 267[26], 293
Weeren, H.O., 274[32], 293
Weinstein, I.B., 105[47], 115
Weissbach, H., 134[13], 145, 182[44], 187, 267[21], 292
Weiss, J.F., 274[32], 293
Wettstein, F.O., 160, 161, 186
White, A.M., 4[9], 27
Wigle, D.T., 267[20], 292
Wilcox, H.B., 215[20], 228
Willems, M., 84[28], 114
Williams, N., 148[2], 184
Williamson, A.R., 77[17], 113, 239[22], 263
Wilson, C., 36[13], 56
Wilson, D.B., 267[16], 292
Wilson, S.H., 46[18], 56
Winget, C.A., 298[7], 302[7], 305[7], 316
Witter, R.F., 65[13], 113
Wool, I.G., 2[2,3], 3[2,5], 6[10], 7[3,12], 9[16], 11[17], 16[10,19], 26, 27, 105[48], 115, 182[42], 187, 291[44], 294, 315[78], 321
Worthington, J., 4[9], 27

Y

Yahara, I., 100[41], 115
Yamazaki, S., 304, 305[31], 318
Yang, P.C., 270[29], 293
Yoshida, A.,]67[18], 292
Younger, J.S., 301[17], 309[47], 316, 319

Z

Zamecnik, P., 157[20], 185
Zimmerman, E.F., 83[20], 113
Zomzely, C.E., 148[5,7], 150[13], 153[14,18], 155[7,18], 16][7,18], 163[7], 166[7,13], 169[7,14], 170[7], 173[14], 175[7], 177[40], 179[40], 183[40], 184, 185, 187
Zubay, G., 130[10], 144
Zucker, W.V., 249[24], 263

SUBJECT INDEX

A

Amino acid incorporation, 51-52, 108-109, 123-125, 128-135, 168-176
Aminoacyl transferases, 18, 20
Aminoacyl-tRNA, 16, 19
Antibodies
 labeled, 223-224
 preparation, 213, 215-216
Antigens
 assay, 213, 216-221
 cell surface, 192, 213
 detection, 213, 216-221
 nascent, 223-224
 preparation, 213-215
 synthesis by isolated gradient fractions, 221-223
 transport, 225-226
AS_{70}, 52-53; see also pH 5 fractions

B

BHK(baby hamster kidney), 30, 62
 amino acid incorporation system, 51-52
 enzyme preparation, 52-53
 I factors, 106-111
 polyribosomes, 40-44, 62-83
 ribosomes, 31, 92-96

C

Cations, 70-73
Cell-free protein synthesis; see Protein synthesis
Cerebral mRNA, 176-182
Cerebral polyribosomes, 159-168
Cerebral ribosomes, 149 ff., 153-155

D

Density gradients; see Sucrose density gradients
Detergents, 70-73
Dextran sulfate, 67-70, 96-100
DOC (deoxycholate) 5, 39-40

E

Elongation factors
 factor T1, assay, 132
 factor T1, preparation, 132
 factor T2, assay, 134
 factor T2, preparation, 132
 factor T1(E), preparation, 141
 Factor T2(E), preparation, 142
Ethanol precipitation of mRNA, 245

F

Factor C; see Initiation factors
Factor D; see Initiation factors
Factor M_1, 277; see also Initiation factors
Factor M_2, see Initiation factors
Factor M_3, 285; see also Initiation factors
Factor T1; see Elongation factors
Factor T2; see Elongation factors
Factor T1(E); see Elongation factors
Factor T2(E); see Elongation factors

H

Hemoglobin mRNA, 231 ff., 238
Hepatic polyribosomes, 44-51; see also Liver
 purification, 50-51

I

I factor
 assays, 108, 109-111
 binding of N-acetyl-L-Phe-tRNA, 109-111
 dissociation from ribosomes, 107
 isolation, 107
 preparation, 106-108
 storage, 108
Initiation factors
 factor C, assay, 136-138, 143
 factor C, purification, 140
 factor C, separation from D, 139
 factor D, assay, 136-138, 143
 factor D, purification, 140-141
 factor D, separation from C, 139
 factor M_1, assay, 280-281
 factor M_1, preparation, 277-279
 factor M_2, assay, 282-285
 factor M_2, preparation, 282
 factor M_3, assay, 288-289
 factor M_3, preparation, 286-288
Interferon, 295 ff.
 assays, 297
 mode of action, 306-315
 preparation, 297-300
 properties, 300-303
 purification, 295-397, 303-306
 unit, 297

L

Liver, 30; see also Hepatic
(Liver, cont'd.)
 amino acid incorporation system, 51-52
 cell lysis, 47-48
 enzyme preparation, 52-53
 isolation and preparation, 46-47
 polyribosomes, 44-51
 ribosomes, 31, 47, 50-51
Liver ribosomes, 31
 isolation, 47
 purification, 50-51

M

Mammalian polyribosomes, 29 ff., 78-83
Medium A, 5
Membrane-bound ribosomes, 92-96, 120-122
Messenger RNA; see mRNA
Microsomes
 cerebral, 149-153
mRNA
 assay, 247-261
 cerebral, 176-182
 concentration by ethanol precipitation, 245
 hemoglobin, 231 ff., 238
 isolation, 233-247
 preparation from ribosomes, 238-245
 spleen, 124
Muscle ribosomes, 1 ff., 2-7

N

N-acetyl-L-Phe-tRNA, 109-111

P

<u>Paramecium</u>, 189 ff.
 crude supernatants, 196-198
 cultures, 193-195
 harvesting, 195-196
 homogenization- 196
 pH 5 fraction, 199
 protein synthesis, 205-212
 ribosomes, 198-199

pH 5 fractions, 53; see also AS_{70}
cerebral, 155-159
mouse liver, 200
paramecium, 199
rat liver, 124
spleen, 122-123
Poly U
binding to ribosomes, 125
binding of N-acetyl-L-Phe-tRNA, 109-111
cerebral system, 172
polyphenylalanine synthesis, 21-26, 101-103, 108-109, 124-125, 128-135
Polyribosome preparation
cations, 70-73
detergents, 70-73
dextran sulfate, 67-70
incubation conditions, 74-78
ionic environment, 33-34
method of cell breakage and homogenization, 32-33
RNase inhibition, 34-35, 41, 66-67
Polyribosomes, 29 ff.
BHK, 40-44, 62-83
cerebral, 159-168
cultured mammalian cells, 40-44, 59 ff., 62
free, 166-168
hepatic, 44-51
isolation, 35-37, 78-83, 119-120
mammalian cells, 29 ff., 78-83
preparation, 119-120, 159-168
purification, 50-51
rabbit reticulocyte, 35-37, 269
tissue culture, 59 ff.
total, 159-166
tumor cells, 44
spleen, 119-120
virus-infected cells, 83-91
Protein synthesis; see also Amino acid incorporation
assay, 16-26
(Protein synthesis, cont'd.)
cerebral, 168-176
directed by TMV RNA, 135
muscle, 16-26
paramecium, 205-212
poly U-directed, 21-26, 101-103, 108-109, 124-125, 128-135
spleen, 117 ff., 122-125
wheat embryo, 127 ff.
Protozoan systems, 189 ff., 226-227

R

Rabbit reticulocyte ribosomes, 31, 35-40, 269
Rabbit reticulocytes, 30
polyribosomes, 35-37, 269-272
ribosomes, 31, 37-40, 236-238
Reticulocytosis
induction 234-236, 268
rabbits, 35, 235, 268
guinea pigs, 235
ducks, 236
mice, 236
Ribosomal subunits
assays, 108-111
muscle, 7-16
preparation, 7-16, 105-106
tissue culture, 105
Ribosomes, 30
cerebral, 149 ff., 153-155
free, 92-96, 120-122
isolation, 198-199
membrane-bound, 92-96, 120-122
muscle, 1 ff., 2-7
paramecium, 198-199
preparation, 2-7, 31-32, 37-40, 120-122, 129-130, 138-139, 153-155, 236-238
properties, 40
purification, 50-51
reticulocyte, 37-40, 236-238, 269-272
spleen, 120-122
"stripped", 183-184
wheat germ, 129-130, 138-139
RNA; see Ribosomal RNA, tRNA, mRNA

RNase inhibitors, 34-35, 41, 66-70

S

Spleen mRNA, 124
Spleen polyribosomes, 119-120
Spleen ribosomes, 120-122
Sucrose density gradients, 8, 53-55, 201-204, 239-244

T

Tissue culture polyribosomes, 59 ff.
Tissue culture systems, 30 ff., 59 ff.
  cell-free studies, 97-105
  polyribosomes, 62-83
  ribosomal subunits, 105-106
  virus-infected, 83-91
Tumor cell polyribosomes, 44
TMV RNA, 135

V

Virus-specific polyribosomes, 83-91

W

Wheat germ ribosomes, 129-130, 138-139